NATURAL NUMBER

The Unified Field

P. S. Williams

Horse Sense Publications

NATURAL NUMBER

www.naturalnumber.com

A Horse Sense Publications book

Horse Sense Publications
PO BOX 80275
Portland, OR 97280

Printed in The United States of America

ISBN: 0615671624
ISBN-13: 978-0615671628

To Sophie
and future generations

TABLE OF CONTENTS

ACKNOWLEDGEMENT

"Pythagoras, son of Mnesarchos, a Samian, who was the first to call this matter by the name of philosophy, assumed as first principles the numbers and the symmetries existing in them, which he calls harmonies, and the elements compounded of both, that are called geometrical."*

*Passage from The Doxographists: Arthur Fairbanks, The First Philosophers of Greece (1898)

INTRODUCTION

In this book I analyze the geometry of the continuous proportions of Pascal's triangle in a spatial/temporal dimensional context. As a means of validating this dimensional number theory, I correlate these findings with the periodic quantum number symmetry of atomic elements. My analysis suggests that quantum states are actually dimensional states of energy governed by the natural number symmetry of Pascal's triangle, as opposed to the closely related constructive interference states of a spherical harmonic waveform (Schrodinger's Equations). I present a compelling series of geometric, mathematical, and symmetric proofs and observations that serve to objectively substantiate my contention that, at its most fundamental quantum levels, energy is best characterized as a quantized multidimensional waveform sustained and propagated by the cooperative spatial/temporal dynamics of the golden ratio and it's negative reciprocal – *the spatial base and temporal anti-base of continuous golden proportion.*

I characterize the continuous proportions of Pascal's triangle as unique dimensions emanating from the same unified singular origin: the point, whose geometric symmetry is represented by the number 1. I show how the mirror symmetry of Pascal's triangle is caused by the geometric relationship between the odd point (1) and the even void (0) that surrounds it. I explain why this mirror symmetry conveys a spatial/temporal relationship between its mirror-twin continuous proportions: For instance, I explain why the twin continuous arithmetic proportion are spatial and temporal linear dimensions, the twin continuous triangular proportion are planar dimensions, and the twin continuous tetrahedral proportion are dimensions that fill both space and time with complementary cuboctahedral spherical close-pack symmetries.

The combination of the continuous equal proportions of point and void produces a continuum of orthogonal continuous proportions – a continuum of virtual spatial/temporal energy dimensions. I have found that I can interpret the dimensional geometry of these continuous proportions using their accompanying complementary modulus-9 symmetry transformations. I show how this modulus-9 symmetry is created by the invariant complementary geometric relationship between the number one and the number zero, point and void.

Ultimately I propose that the continuous golden proportion of Pascal's triangle, the Fibonacci series, is governed by the complementary-cuboctahedral rotational symmetry transformations of a spatial-temporal dipole inversion field. And, I associate this complementary-cuboctahedral dipole field symmetry with that of the electron, proton, and neutron. This then leads to a reinterpretation of the periodic quantum number symmetry of atomic elements in terms of the dimensional number symmetry of Pascal's triangle, where quantum states are equated with the dimensional states of energy.

Additionally, I have characterized the continuous triangular proportions that govern all forms of Pythagorean Triples and their corresponding prime number symmetry fields. I have found that the continuous proportions of the in-circle and ex-circles of the near isosceles class of triples, beginning with the 3-4-5 triangle, correspond with orthogonal linear and triangular continuous proportions of Pascal's Triangle, and that the continuous proportions of triples diagonals are composed of interleaved continuous prime proportions. Further, I have found that these interleaved continuous prime proportions have orthogonal spatial and temporal components reflective of their formative orthogonal triangular and linear dimensions. I've determined that these continuous prime proportions are governed by modulus-9 complementary symmetry transformations between primary and secondary prime chromo-numeric phases. Ultimately, I propose that the dimensional symmetry of Pythagorean Triples and their associated continuous triangular/linear (spatial/temporal) proportions is a strong candidate for the symmetry that mediates electromagnetic phenomena.

Essentially I believe that I've uncovered the geometric link between continuous proportions and spatial/temporal dimensional symmetry by leveraging the numeric invariance of modulus-9 to reveal the complementary symmetry transformations that accompany each continuous proportion. In so doing, I've adopted the Pythagorean premise that everything is in accord with number. I believe that when science seeks to answer fundamental questions like what a proton is or why atomic elements have the periodic symmetry they do, science is really seeking absolute truths, axiomatic comprehension of why everything is as it is. My thinking is that this requires comprehension of the dimensional symmetry of number itself, because I am convinced that the order within number is inseparable from the order of the universe. Dimensional number theory suggests that everything is connected to the same rational geometric *first cause* – the force of spatial/temporal dimensional symmetry that creates the symmetry of the electron and proton and all that arises there from. In the following chapters I will attempt to establish a dimensional number theory that integrates symmetry, geometry, physics, and even philosophy as components of a unified transcendent absolute paradigm.

Understanding this content may seem challenging. However, the mathematics of continuous proportion is elegantly simple and is accessible to anyone possessing basic math and science skills with some effort. The subject matter is not abstract like many other forms of mathematics; this enables geometric evidence to be provided so the reader may understand without needing to decipher most of the equations (they are provided mainly for the more fluent math reader). Actually, the application of modulus-9 levels the playing field making the subject matter visually and intuitively accessible to a wide audience.

Beyond Relativism

> ... some things are for the sake of something else (relatives), and ... other things are the ends to which the former class subserve (absolutes).
>
> Socrates [1]

2.1 Relative Realities

Current quantum theory leverages Schrödinger's three dimensional harmonic oscillator wave equations to define the quantum numbers used to distinguish the periodic energy states occupied by subatomic particles within atomic elements. Yet this model can't accurately predict electron energy levels beyond Hydrogen. Nor does it predict the spin up/down state splitting of these energy levels. Schrödinger himself expressed reservations with the model in his famous "cat paradox" paper:

> ... the so-called wave function or psi-function... is an abstract, unintuitive mathematical construct ...
>
> E. Schrödinger [2]

The process of model building to explain experimental results is like knowing the effects and explaining them with a hypothetical causal analogy to some known cause-effect relationship, such as the well-characterized physics of harmonic oscillators in the case of quantum mechanics. When such a model is followed by a series of refinements in the causal analogy, as is the case with quantum mechanics, it brings into question the validity of the initial hypothesis. If cause and effect were truly understood axiomatically, by which I mean understood in terms of self-evident natural laws, then no further refinement would be necessary.

The incremental successes of the succeeding refinements of the quantum mechanical model seems to have lead many physicists down the path of essentially learning to live with the model whether or not it is well reasoned from first principles. And further, proponents of this theory don't appear to mind as long as it continues to give *predictions that agree with experiment*, as Nobel laureate Richard Feynman puts it:

> It's a problem that physicists have learned to deal with: They've learned to realize that whether they like a theory or they don't like a theory is *not* the essential question. Rather, it is whether or not the theory gives predictions that agree with experiment. It is not a question of whether a theory is philosophically delightful, or easy to understand, or perfectly reasonable from the point of view of common sense. The theory of quantum electrodynamics describes Nature as absurd from the point of view of common sense.
>
> R. Feynman [3]

I have difficultly abandoning *common sense* and accepting *Nature as absurd*. To me it is illogical to presume that the dimensional symmetry that governs the nature of matter does not also govern the nature of all that arises from it including me. If I'm asking what governs atoms, I'm really asking what are the first principles that govern Nature. This has led me to search for a holistic answer to why everything is as it-is.

As I began my studies I asked myself whether the answer to this broader question could ever be found in a compartmentalized laboratory experiment, and I questioned whether current paradigms in the scientific world were not in fact subtly hindering the search for the absolute truths of why things are as they are. For me the reductionist method of dissecting reality and constructing models based on experimental results was wedged against the hard barrier of its own inherent limitations. I was seeking to move beyond a relativistic interpretation of reality, which I view as inherently subjective since it is always relative to a given observer at some differentiated point in space-time who views effects, and discerns causes. I was looking for a unified absolute perspective from which the truth of why everything is as it is would become self-evident.

2.2 ABSOLUTE AXIOMS

I decided to leverage quantum physics to validate a theory of everything founded upon observations of the axiomatic first principles of continuous proportions. As Sterling Professor Emeritus of Physics at Yale University R.K. Adair puts it in his book *The Great Design*:

> ... there is a longstanding interest in providing an axiomatic basis for quantum field theory (as Euclidean geometry is axiomatic) ...
>
> R.K. Adair [4]

As I probed these matters it became apparent that the current models used to define dimensions are inconsistent with the axiomatic principles of Euclidean geometry developed from the Platonic and Pythagorean schools. In modern science linear dimensions are typically combined to form a three dimensional construct using the Cartesian coordinate system (x, y, z), which may alternatively be converted to the Spherical-Polar coordinate system (r, θ, ϕ) used in Schrödinger's spherical harmonic oscillator equations. I found the presumed 3-dimensional limit on spatial dimensions and the manner in which dimensions are formed in this model at odds with the geometric axiom that dimensions are formed by the combination of points: two points define a line, three a plane, and four a solid. Further, I was unconvinced that a spherical-polar coordinate system was sufficient to model the geometric complexity of observable phenomena. I turned instead to the dimensional geometry of continuous proportion.

Since my own education was deficient in the study of continuous proportion, I set out to remedy this by studying ancient Greek science and philosophy. I chose this route because I found very little regarding continuous proportion in modern texts. This appeared to me to be a glaring omission since the study of numeric proportion was so important to Greek mathematicians.

> He who knows must have learned from another, or have found his knowledge by himself. The science that you learn from another, is as you might say, exterior; what you find by yourself, belongs to ourselves individually. To find without seeking is something difficult and rare; to find what one is seeking is commodious and easy; to ignore, and seek what you ignore, is impossible.
>
> Archytas[*] [5]

I wondered if modern ignorance of continuous proportion was the blind spot that was preventing quantum physics from moving past the art of chasing *statistical particle shadows* (as I saw them). So following Heisenberg's sound advice I began to study Greek number theory:

> … the concept of the atomic structure of matter dates back to the philosophy of Antiquity, and was suggested by Greek philosophers as a daring hypothesis 2,500 years ago. Anybody who desires to understand something of modern atomic theory, will do well to study the history of the concepts of the atom …
>
> W. Heisenberg [6]

---

[*] Archytas of Tarentum was student of Philolaus and personal friend of Plato, 4th century BCE.

I found that Pythagorean thought in particular meshed well both with rational and intuitive notions of reality. I think that the key to understanding reality is to obtain the perspective wherefrom natural laws are self-evident. This involves both looking in the right place and observing from the proper perspective. To do this I had first to step outside the modern paradigm where scientific, philosophical, and metaphysical notions of reality are compartmentalized, and understand that these different modes of inquiry are ultimately directed toward interpreting the same one-reality. I felt that I could never attain the perspective necessary to observe natural laws unless my perspective was devoid as possible of any modern preconceptions that would bias my observations. Pythagoras provided me with the foundation of such a perspective since in his day these artificial divisions didn't exist.

Pythagoras of Samos (c. 582 -500 BCE) lived in an age where wandering sages traveled among the many city-states of the near east. He was a contemporary of the Buddha (c. 563-483 BCE), Confucius (c. 551-478 BCE) and a host of other enlightened champions of wisdom. This was an age of great diffusion and cross-pollination of philosophic thought, and it marks the end of a period characterized by the development of The New City States (c. 800-500 BCE) where Aryan warrior herdsmen dominated the region from Greece to India.[7] It was the beginning of what Joseph Campbell describes as The Age of Great Classics (c. 500 BCE to 500 CE) where Zoroastrian, Hebrew, Christian, Gnostic, and Manichaean canons were defined.[8] German philosopher Karl Jaspers referred to the time of Pythagoras (centered about 500 BCE) as the *Achsenzeit* (Axis-time) from which the term Axial Age was coined – a time when the foundations of modern thought were laid.[9]

As a wandering sage and student of many of the influential mystery schools of his age, Pythagoras' wisdom was a synthesis and elaboration of the thought forms of his time, thought forms that were the nucleating elements for nearly all of the worlds great religions. Iamblichus summarized the scope of Pythagoras' wisdom in his biographical account of the sage's life as follows:

> ... his divine philosophy and worship was compound, having learned much from the Orphic followers, but much also from the Egyptian priests, the Chaldeans and Magi, the mysteries of Eleusis, Imbrus, Samothracia, and Delos, and even the Celtic and Iberian.
>
> Iamblichus[*] [10]

Only glimmers of Pythagoras' most closely guarded teachings survive since he left no written records. Through oral tradition and the writings of following generations a few

---

[*] Iamblichus was an influential Neo-platonic philosopher and student of Porphyry, c. 250 – 325 CE.

core tenets remain intact that provide vital clues to help unravel the mystery of his unified number field theory. For instance, the Pythagorean's emphasis on the triangular number 10, the *Tetraktys*[11], suggests an understanding of the significance of modulus-9 analog number symmetry. However, I was looking for more than a few clues from Pythagoras – I was looking for a unified absolute perspective of reality to replace the menagerie of divisive perspectives in current thinking. I wanted to see the world through the eyes of a Pythagorean. I sought to learn the way a Pythagorean learned:

> The way is this: every diagram, and system of number, and every combination of harmony, and the agreement of the revolution of the stars must be made manifest as one through all to him who learns in the proper way, and will be made manifest if, as we say, a man learns aright by keeping his gaze on unity; for it will be manifest to us, as we reflect, that there is one bond naturally uniting all these things ...
>
> Plato [12]

With ancient philosophical fragments as my guide, I set my sight on a higher level of understanding. My quest for the *unknowable* truth of the ALL began in earnest. Through meditation I purged my preconceptions, *emptying my cup* that I may receive the absolute wisdom of number – the true light with the capacity to illuminate that that hides in the darkest shadows of ignorance. I found that by seeking to understand ancient wisdom, a doorway to new thinking opened to me. Ancient wisdom provided me with a seed thought, and my mind was made ready like fertile ground for those thoughts to germinate. Perhaps more than anything the ancients had taught me a method of reasoning:

> ... the ancients, who were our betters and nearer the gods than we are, handed down the tradition, that whatever things are said to be are composed of one and many, and have the finite, and infinite implanted in them: seeing, then, that such is the order of the world, we too ought in every enquiry to begin by laying down one idea of that which is the subject of enquiry; this unity we shall find in everything.
>
> Socrates [13]

> With Reason (*Logos*), not with hands, did the World-maker make the universal World; so that thou thus shouldst think of Him as everywhere and ever-being, the Author of all things, and One and Only, who by His Will all beings hath created. The Oneness being Source and Root of all, is in all things as Root and Source. The Oneness then being Source, containeth every number, but is contained by none; engendereth every number, but is engendered by no other one.
>
> Hermes Trismegistus [14]

Socrates relates the ancient method of reasoning and Hermes provides an example pertinent to the topic at hand. Thrice Great Hermes* in no uncertain terms states that the universe is governed by reason and that its source is the number one. The Greeks generally associated the number one with the point (the *monad*). I concluded that if I considered the number one to represent the geometry of the point and the singular dimension, then the point itself would represent the symmetry of a singularity, such as the one proposed to have preceded the "*Big Bang*" at the dawn of the universe or the one associated with the "*black hole*" which is thought to be at the heart of our own galaxy. I would accept this idea with the caveat that such a state could be possible only if it were a singular energy state, a singular waveform, versus some exploding superheated quark-gluon plasma.

For me the latter is a very important distinction because I can comprehend a singular energy state where every iota of energy in the entire universe is in the same undifferentiated state – the waveform of a singular spatial-temporal inversion center, I reasoned. This interpretation led me to consider the dimensional symmetry of energy as one that is senior to matter since I agree with Plato that *a thing governing is senior to a thing governed.*[15] And I would add that I believe that a higher level of symmetry governs a lower in the sense that lower symmetries exist within higher symmetries but not vice versa.† Thus when I consider my own symmetry, that is the symmetry of my own body, I recall the Pythagorean notion that *man is a microcosm,*[16] and this causes me to look for my own symmetry within the symmetry of the singularity. Immediately I think of the Golden Ratio since it *plays a dominating part in the proportions of the human body.*[17] And as golden proportion is most symmetric relative to other forms of proportion, it seemed reasonable to me that it would govern the singular energy state.

I consider Pythagoras to be one of the first unified field theorists. Long ago he suggested that one should seek to understand the ordering symmetry of number itself — the link between number and spatial/temporal geometric symmetry, I reasoned. For when I recalled the Pythagorean notion that *all things accord in number,*[18] I inferred that the symmetry of energy must be in accord with some form of spatial-temporal dimensional number symmetry. Further, I reasoned that energy dimensions must govern atomic matter

---

*Hermes Trismegistus: revered in the Hellenistic period as a mythological synthesis of the wisdom attributed to the Greek messenger of the gods, Hermes, and the Egyptian deity Thoth, since it was widely held that they were the same deity owing to their many parallel mythic roles.

†For instance 2-fold is a subset of 4-fold but not vice versa. So 2-fold can come from 4-fold, but 4-fold cannot come from 2-fold.

– matter being composed of energy implies energy is senior to matter dimensionally.* Therefore, I thought it reasonable to expect the periodic symmetry of atomic elements to be reflective of this governing spatial/temporal energy symmetry.

This reasoning led me to the conclusion that the dimensional symmetry of number arises from the symmetry of the singular spatial/temporal energy state, which must be in harmony with golden proportion. In one of many epiphanies, I surmised that the quantum states of matter are spatial/temporal energy symmetry states. And, I reasoned that I should look for a numeric symmetry field associated with golden proportion where the number one is the origin of every dimension – Knowing ultimately that this field must somehow equate with the quantum states of atomic matter and reveal symmetries consistent with those observed by particle physicists. The only numeric symmetry field that meets these prerequisites is "Pascal's" triangle†.

Ultimately I find that when the geometry of Pascal's triangle is interpreted in a spatial-temporal dimensional context and analyzed with complementary modulus-9 symmetry, super-symmetric spatial-temporal cuboctahedral dipole fields are revealed that provide an axiomatic means of understanding the natural order empirically characterized by quantum physics.

---

*Energy dimensions, being established by a preordained virtual mathematical symmetry, exist prior to their expression within matter in the form of filled energy states.

†The earliest documented study of Pascal's triangle is attributed to Pingala who gave it the name *Meruprastāra* (the expanding mountain of jewels). Pingala documented his work on Vedic Meter in the Chandashāstra (c. 200 BCE), in which he characterized the possible rhythmic combinations of long (*guru*) and short (*laghu*) syllables in Sanskrit prose using binomial expansions (*prastāras*).[19] [20]

CHAPTER CITATIONS

---

[1] Plato *Philebus* trans. Jowett, B. http://classics.mit.edu//Plato/philebus.html
[2] Schrödinger, Erwin 1935 *Die Naturwissenschaften* trans. Trimmer, J. D. 1980 *The Present Situation In Quantum Mechanics: A Translation of* Schrödinger's *"Cat Paradox Paper"* Proceedings of the American Philosophical Society, 124, 323-38
[3] Feynman, Richard 1985 *QED The Strange Theory of Light and Matter* Princeton University Press (NJ) p 10
[4] Adair, R. 1987 *The Great Design – Particles, Fields, and Creation* Oxford University Press (NewYork) p 209
[5] Archytas trans. Guthrie, K.S. 1988 *The Pythagorean Sourcebook and Library* Phanes Press (USA) pp 183-4
[6] Heisenberg, W. 1953 *Nuclear Physics* Philosophical Library (New York) p 1
[7] Campbell, Joseph 1962 *Oriental Mythology* Penguin Books (USA) pp 246-250
[8] Campbell, Joseph 1962 *Oriental Mythology* Penguin Books (USA) p 288
[9] Jaspers, Karl 1953 *The Origin and Goal of History* Yale University Press (New Haven)
[10] Iamblichus trans. Guthrie, K.S. 1988 *The Life of Pythagoras - The Pythagorean Sourcebook and Library* Phanes Press (USA) p 95
[11] Guthrie, K.S. 1988 *The Pythagorean Sourcebook and Library* Phanes Press (USA) p 28
[12] Plato *Epinomus* trans. Lamb, W.R.M. http://www.ac-nice.fr/philo/textes/Plato-Works/29-Epinomis.htm
[13] Plato *Philebus* trans. Jowett, B. http://classics.mit.edu//Plato/philebus.html
[14] Mead, G.R.S. 1906 *Thrice-Greatest Hermes Vol. 2* p 85, pp 90-91
[15] Plato *Epinomus* trans. Lamb, W.R.M. http://www.ac-nice.fr/philo/textes/Plato-Works/29-Epinomis.htm
[16] Guthrie, K.S. 1988 *The Pythagorean Sourcebook and Library* Phanes Press (USA) p 31
[17] Ghyka, Matila 1977 *The Geometry of Art and Life* Dover Publications (New York) p 16
[18] Guthrie, K.S. 1988 *The Pythagorean Sourcebook and Library* Phanes Press (USA) p 97
[19] Hall, R.W. 2007 *Math for Poets* p.9 (Internet Archive)
[20] Kulkarni, A. 2008 *Recursion and Combinatorial Mathematics in the Chandashāstra* (Internet Archive)

Dimensional Perspectives

> The universe is one, and it began to come into being from the centre, and from the centre upwards at the same intervals (of distance) as those below. For the parts above from the centre are in inverse relationship to those below; for the centre is to what is below as it is to what is above, and so with all the rest; for both stand in the same relationship to the centre, except in so far as their positions are reversed.
>
> Philolaus* [21]

The ancients had no modern telescopes to inform them that the cosmos is expanding and thus must originate from a singularity – a singular center of inversion. Yet through observation of the first principles of number and geometry they concluded that it did. What are the first principles that led to this conclusion? What is the link between number and geometry that so fascinated the Pythagoreans?

> Everything has been organized by the monad, because it contains everything potentially: for even if they are not yet actual, nevertheless the monad holds seminally the principles which are within all numbers, including those which are within the dyad. For the monad is even and odd and even-odd, linear and plane and solid (cubical and spherical and in the form of pyramids from those with four angles to those with an indefinite number of angles)...
>
> Iamblichus [22]

> If as many numbers as we please beginning from an unit be in continued proportion, and the number after the unit be square, all the rest will also be square. And if the number after the unit be cube, all the rest will also be cube.
>
> Euclid* [23]

---

*Philolaus of Croton was educated by Lysis who was a student of Pythagoras, 5th century BCE. This quote is from a work entitled *Bacchae* sometimes attributed to him.

Both Iamblichus and Euclid suggest that we should consider the number one (*monad* or *unit*) to have the flexibility to originate any spatial form. Euclid further suggests that continuous proportions are characteristic of these geometric forms. Given that the study of numeric proportion and its relationship to geometry was an integral part of the science of the Pythagoreans and the Greeks that followed, I concluded that studying continuous proportions might help me understand the first principles of dimensional symmetry.

Like everyone else I was accustomed to believing that there are three dimensions of space and one dimension of time. Each spatial direction is generally considered to be linear and measured with Cartesian (x, y, z) coordinates. We generally traverse space in a straight line from point A (x1, y1, z1) to point B (x2, y2, z2) and observe time coupled with space as velocity and acceleration. These notions of space and time are effectively linear and are not particularly suggestive of any underlying symmetry.

Certainly a linear perspective is useful for traversing empty space, but how useful is it for filling space/time with energy and matter? How can such a dimensional model explain the energy symmetry states that electrons occupy in matter? Quantum theory characterizes the symmetry of self-ordered energy states within matter (effects) but it doesn't convey any causal source for said effects. It seems logical to presume that the answer to this causal mystery is linked to the dimensional symmetry of energy. Where else could this symmetry come from but the spatial-temporal dimensions of energy itself, I thought.

I began to think that our individual relativistic perspectives might be deceptive since we interpret our observations relative to our own positions in space-time versus its singular origin. Instead of this, I wondered, what if I were able to adopt the perspective of this singular origin? Might I in so doing obtain an absolute perspective and observe a unified dimensional symmetry field emanating from this singular origin? To begin I had to purge my preconceptions in order to be informed solely by the perspective of this singular origin.

Initially I found this to be quite challenging since I had no means of defining a perspective. Then I decided to define a perspective to be a certain vantage of the point characterized by a continuous proportion. As *the origin of all things*, the point then would be the origin of all dimensional perspectives, *just as a point is the beginning of a line, a line of a surface, and a surface of a solid.*[24] As all dimensions are constructed from points, it follows then that continuous whole number proportions that originate from the number one should define all dimensional perspectives that originate there from. At this juncture in my thinking the continuous proportions of Pascal's Triangle (*Meruprastāra*) became the

---

[*]Euclid's Elements Book Nine: Proposition Nine

foundation of my study of dimension number theory. I found that by considering the number one the geometric origin of the continuous proportions of Pascal's Triangle, I could from this vantage observe a continuum of dimensional perspectives emanating from the point and their governing complementary cuboctahedral super-symmetry.

There is no historic evidence that the Pythagoreans were aware of Pascal's Triangle. Yet it is hard to imagine that they weren't, given that they were students of continuous proportions and understood *Figurate Number*, which is a two dimensional representation using polygons of the various continuous proportions of Pascal's Triangle. In this work I will use the term "dimensional number" instead of the terms figurate or polygonal number in reference to the dimensionally ordered continuous proportions that comprise Pascal's Triangle.

Modern convention designates the point as the 0th dimension (0D) since it is un-manifest or has no physical size; succeeding dimensions are designated as 1D or scalar (*x*), 2D or planar (*x*, *y*), and 3D or volumetric (*x*, *y*, *z*). I suggest instead that the continuous geometric proportions of Pascal's Triangle should be used to define dimensions. A version from his writings is provided as Figure 1. Each row (*Rangs paralleles*) and each column (*Rangs perpendiculaires*) of Pascal's Triangle has continuous whole number proportions beginning with the continuous equal proportions of the singular point (first column and row— all ones), followed by the continuous arithmetic proportions of the line, then the continuous triangular proportions of the plane, then the continuous tetrahedral proportions of volume, and beyond a continuum of increasingly complex dimensions.

By equating the singular point with the number 1, it follows that all dimensions constructed of points should all have continuous whole number proportion. These dimensions can be distinguished or differentiated from the dimension of their common singular origin by their unique continuous proportion whose geometry can be established by the first magnitude in the proportion beyond the origin: it takes two points to define a line (the linear dimension), three points to define a plane (the triangular dimension), and four point to define a volume (the tetrahedral dimension). Higher dimensions are difficult to visualize since they fold back in opposition to previous dimensions (introvert) subsequent to the tetrahedral dimension, and this creates an analog series of extroverted and introverted dimensions. To grasp how this works it is critical to understand the first four dimensions in detail because it is no great task to observe the rules that govern their evolving geometry and number symmetry. Later I'll extend these rules to higher dimensional perspectives and confirm this interpretation using modulus-9 number symmetry.

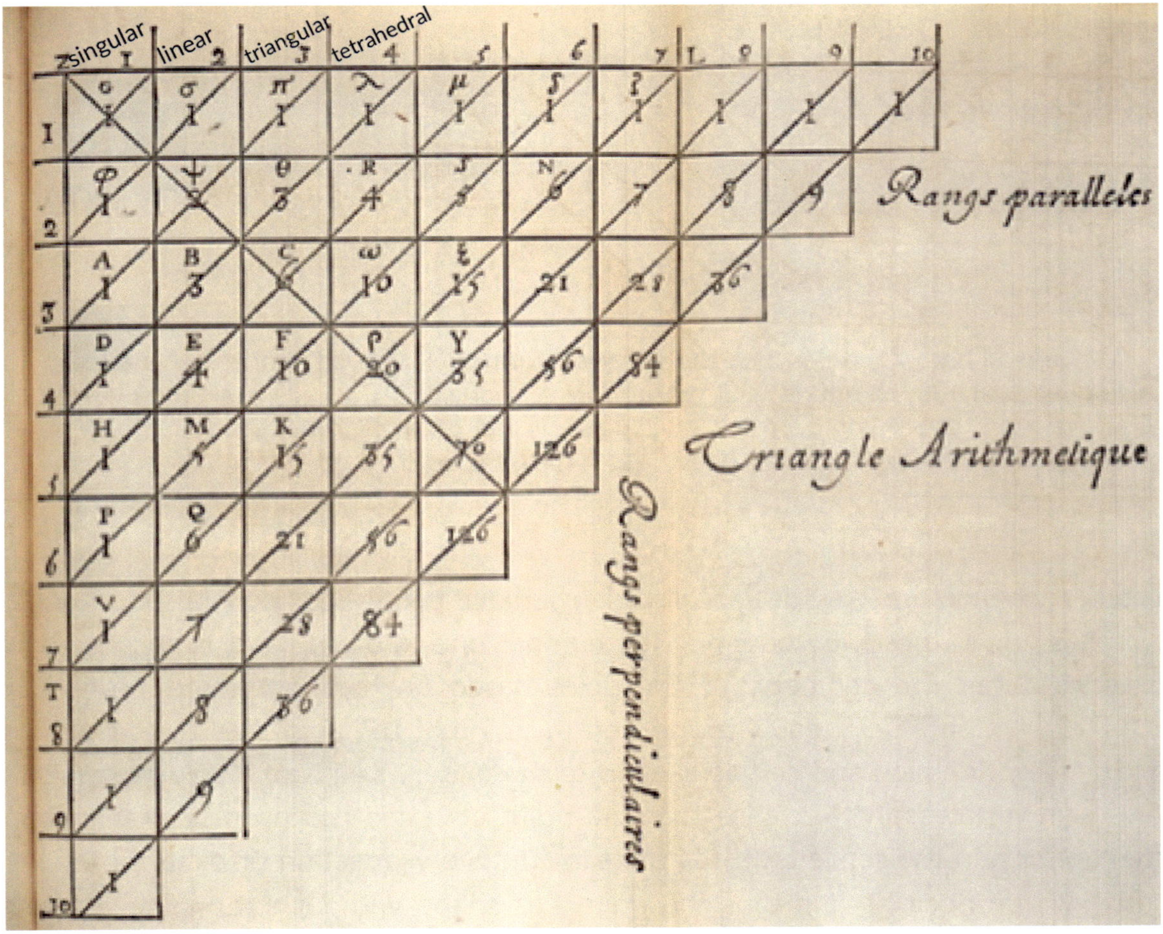

Figure 1. Pascal's Triangle from Blaise Pascal's writings (Triangulo de Pascal en el escrito original de Pascal[25]). Rows are designated *rangs paralleles* and columns *rangs perpendiculaires*. Labels have been added for the singular, linear, triangular, and tetrahedral continuous proportions. Each cell is equal to the sum of the preceding adjacent cells above (same column) and left (same row).

As is evident, both the columns and rows of Pascal's Triangle have the same continuous proportions. Through geometric observations in this chapter I will show that these dual column and row perspectives (dimensions) mirror each other, and that as a result they have a negative reciprocal geometric relationship. Although instead of the negative sign, direction is distinguished by handedness in Pascal's Triangle: envision a number line where to the right of zero is positive and to the left negative without the negative signs on the left. Instead of the reciprocal function, inverse is distinguished by geometric form orientation: envision the two equilateral triangles that compose the "Star of David" motif for example, or the two tetrahedrons that form the star tetrahedron. Following this line of reasoning, I have designated the continuous proportions of Pascal's Triangle's columns as spatial dimensions and its rows as temporal dimensions because, like space and time, they have a negative reciprocal geometric relationship.

3.1 The Origin of Perspective

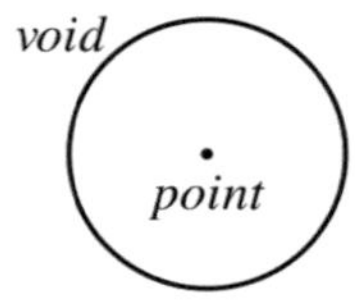

> Why do all natural bodies take the spherical form? Is it, as said Archytas, because in the natural movement is the proportion of equality? For everything moves in proportion; and this proportion of equality is the only one which, when it occurs, produces circles and spheres, because it returns on itself.
>
> Archytas [26]

The first column and row of Pascal's Triangle have continuous equal proportions since every cell is equal to one. By equating the number one to the point it is possible to draw conclusions about the geometry of the point based on the continuous proportions it participates in. Continuous equal proportion means that the point has the same singular length in every direction and can be likened to the unitary radius of a volume-less sphere. So, although formless in the sense that the point or origin has no measurable size, the point is uniform in the sense that it is the same in every direction or dimension. This is a key observation because it means the point has shape without size, namely circular in plane and spherical in volume. The point participates in all the continuous proportions of Pascal's Triangle as their common origin. Therefore it is not only present in every dimension (*omnipresent*), it is the same in every dimension in the sense that it is continuously equal to itself from its singular perspective. However, the point harbors the capacity to originate a continuum of geometric symmetries just as one might inscribe a progression of geometries within a sphere such as a line, an equilateral triangle, a regular tetrahedron, a cube, an octahedron, and so on.

> All creatures under heaven are born from being; Being is born from nonbeing.
>
> Lao Tzu, Tao Te Ching [27]

The point is not the only continuously equal proportion of Pascal's Triangle. For a point to be observable it must be contrasted by an antithetical symmetry element, for without an antithesis its existence would be indeterminable. And, if in the beginning that point is a thing and the only thing, the source of all geometric forms or all-that-is (*being*), what else could enable one to observe its solitary existence other than all-that-isn't, nothingness, void (*non-being*). So *in the beginning* the finite point exists in a solitary state surrounded by void — a singular spatial/temporal state within an infinite expanse of nothingness.

What then distinguishes point from void? The geometry of Pascal's Triangle suggests that the infinite void that subtends finite point is as a tangent is to a radius because when point and void are summed the resulting continuous proportion is always at right angle to the continuous proportion that precedes it. The point is continuously equal to itself and therefore spherical. The continuous proportions of Pascal's Triangle radiate from the point, hence the point is said to have the perspective of a radius. The formless circumferential void that surrounds the point is also continuously equal to itself. But in contrast to the point, the void is infinite and always tangent to the perspective of the point. The juxtaposition of these formative dimensional symmetry elements is shown in Figure 2. The geometric state shown in this figure corresponds to the *first interval* shown in Figure 3. This link between number and geometry (1 = point and 0 = void) sets the foundation for interpreting the continuous proportions of Pascal's Triangle in a dimensional context. The rationale for the details of Figure 2 will fall into place as the coming analysis unfolds. For now the key takeaway is that point and void are always at right angle to one another.

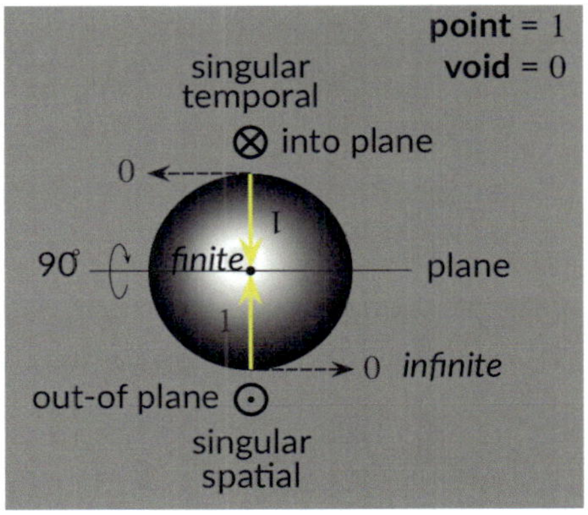

Figure 2. Point (1) and Void (0) shown rotated 90° relative to the plane of the page (condition will be defined as into and out-of the page plane going forward).

When Pascal's Triangle is interpreted in a dimensional context, the initial number one represents the germinal state of the point – consider this state the dimensional origin of space-time. This singular state is shown in Figure 3 as part of the *first interval* together with the orthogonal void that initiates spatial and temporal continuous linear proportion. The void that surrounds the point is implicitly represented by the

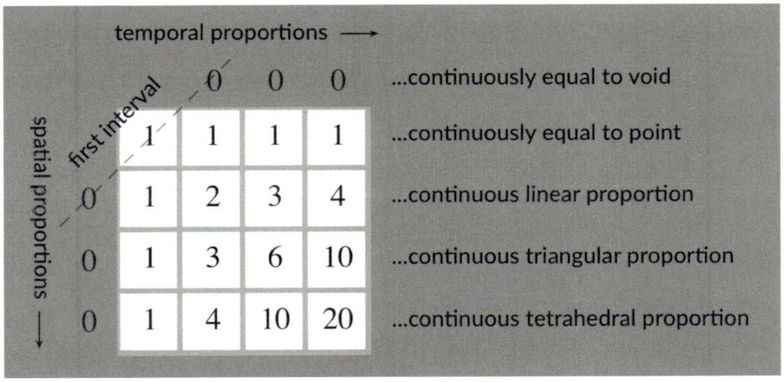

Figure 3. The continuous dimensional proportions of Pascal's triangle begin with the contrasting continuous equal proportions of the tangential-void and the radial-point. Temporal proportion occupy rows and spatial proportion occupy columns.

absence of number that precedes both singular spatial and temporal continuous equal proportions. In other words the column and row that precede the column and row filled with ones are both implicitly filled with zeros, they are just not shown in conventional versions of Pascal's Triangle. The row that is continuously equal to one is generated through summation with the preceding row, which is continuously equal to zero (shown also in Figure 5). The continuous proportions of the columns mirror those of the rows.

Spatial proportions are defined by temporal increments and temporal proportions are defined by spatial increments, as shown in Figure 4. Hence, the direction of the differentiation of time is the space-filling direction, and the direction of the differentiation of space is the time-filling direction. Notice that each number (each table cell or state) of a given continuous proportion has both spatial and temporal components, since each number satisfies both spatial and temporal continuous proportions. In one example, the number three satisfies both linear spatial and triangular temporal proportions and vise versa. In another example, the number ten can be either spatially triangular or temporally tetrahedral or vise versa. And in the case of the singular origin, the number one is both spatially and temporally singular and participates in both spatial and temporal proportions that are continuously equal to one.

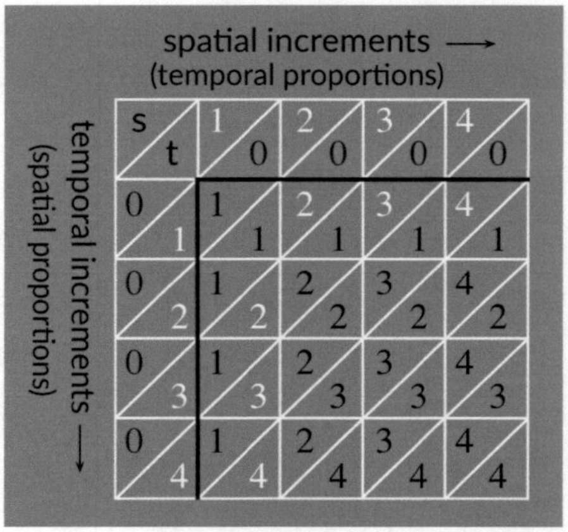

Figure 4. The spatial and temporal increments corresponding to the continuous proportions of Figure 3 highlight the s/t dimensional duality of each differentiated numeric state within Pascal's triangle.

Since the tangential perspective of the void is orthogonal to the radial perspective of the point when the two are summed (0+1), the point remains itself yet becomes constrained to the perspective of that with which it has merged. Therefore, the combined perspectives of void and point constrain the continuous proportions of succeeding columns or rows of Pascal's Triangle to be at right angle (orthogonal) to one another, as indicated in Figure 5. As affirmation, I will show geometrically with Figure 7 that the perspective of each dimension is orthogonal to the perspective of the dimension that precedes it. I will also show with Figure 7 that summing the continuous proportion of a given dimension forms the continuous proportion of the next dimensional perspective, an observation that corresponds to the notion that summation requires orthogonal perspective so that the continuous proportion of the preceding dimension is observable.

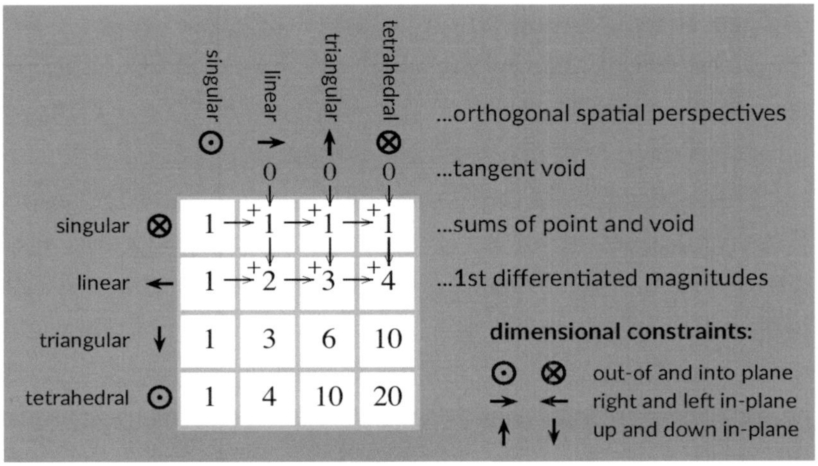

Figure 5. The first four spatial and temporal dimensional perspectives (continuous proportions) are shown ordered by orthogonal dimensional constraints of point (1) and void (0). When void is summed with point it geometrically constrains consecutive continuous proportions to be at right angle to one another – consider point and void as orthogonal vectors as an analogy. The symbols used going forward to define the resulting six possible orthogonal dimensional constraint orientations are shown in the lower right of the figure.

3.2 SPATIAL AND TEMPORAL PERSPECTIVES

> ...there are some who say that there is a right and left of the heavens, as, for instance, those that are called Pythagoreans (for such is their doctrine).
>
> Aristotle [28]

The designation of spatial and temporal orientation is based on *chirality* (handedness). Observation of natural phenomena suggests nature is ordered with chiral mirror symmetry and for the most part nature exhibits no obvious chiral exclusion of either right-hand (counterclockwise) or left-hand (clockwise) phenomenology. Yet in 1956 Richard Dalitz found evidence the *weak force* discriminates between right-handed and left-handed particle interactions — confirmed by C. N. Yang and T. D. Lee the same year.[29] More recent experiments have confirmed that *the W boson* (force) *couples only to left-handed fermions* (electrons) *and right-handed anti-fermions,*[30] consistent with the Standard Model of quantum mechanics; this has led some physicists to state that *God is left-handed.*[31]

Defining handedness requires a consistent frame of reference. To maintain consistency scientists use what's called the *right-hand rule*. Using your right-hand you define the natural thumb-up direction as the axial reference direction, then the direction your fingers curl becomes the right-hand direction of rotation. If you consider looking at the tip of your thumb like looking at the face of a clock, the curl of the fingers on your right hand is counter-clockwise. Alternatively, consider the direction of your right-thumb as the direction of the North Pole and the direction your fingers curl is the direction the Earth rotates on its axis (opposite the direction the sun appears to travel). If you consider the direction your right-fingers curl as the direction of electrical current in a circular winding of wire, then the magnetic field generated by this current is in the direction of your thumb. So when scientists say the electron is left-handed they mean that if you define your left-thumb as the direction of the electron's own axis, then it spins on its own axis in the direction your left-fingers curl (clockwise).

That being said, an axiomatic first principles approach has no *a priori* means of determining whether by nature spatial perspectives are truly right-handed and temporal left or vice versa. So to begin, **spatial** perspectives are arbitrarily defined as the continuous proportions occupying the columns of Pascal's Triangle; the continuous proportions in rows are then designated as **temporal** perspectives. Spatial and temporal continuous proportions mirror one another. This mirroring puts temporal perspectives on the left of the diagonal mirror line and spatial on the right. It remains to be shown how this selection in handedness corresponds to the right-hand-rule. So for now the reader should simply be aware that spatial and temporal perspectives have opposite handedness because of spatial-temporal mirror symmetry. The designation of which is which can only be made arbitrarily for now, knowing that they can easily be reversed if there is a compelling reason to do so.

Next I'll use Figure 6 to show one reason I've concluded the spatial and temporal perspectives mirror one another. Figure 6 examines the intersection of the first four spatial and temporal continuous proportions along the diagonal mirror symmetry line of Pascal's Triangle. Inspection of the geometric species of this diagonal is informative because it enables one to determine the relative orientation of spatial and temporal continuous proportions (dimensional perspectives). The bisecting diagonal of Pascal's Triangle is a line of mirror symmetry. Numbers along this diagonal are composed of equivalently differentiated spatial and temporal species. For instance the triangular number 6 is composed of the spatial 3 and the temporal 3. Numbers along the diagonal mirror line, like the number 6, contain the geometries of both their spatial and temporal constituents. Examination of these constituents suggests that the geometric orientation spatial and temporal continuous proportions are opposite.

The orientations of the first four spatial and temporal dimensional perspectives are indicated by circle-dot and circle-x for out-of and into the plane of the page, and by arrows for right, left, up, and down in the plane of the page.* I refer to these orientations

*The geometric analysis in this chapter first examines the anti parallel orientation of spatial and temporal constraints, and then the analysis demonstrates the relative geometric orientation of the continuous arithmetic (linear), triangular, and tetrahedral proportions. These observations are confirmed later in the book by analyzing the complementary cubic rotational super-symmetry of the continuous geometric proportions of the diagonals of Pascal's Triangle confirmed through modulus-9 analysis. The orientation of singular perspectives is inferred by this same 6-fold pattern of rotation between orthogonal dimensional constraints.

as directions of constraint. These natural directions of constraint that arise from the merger of point and void are like lines of sight that distinguish the dimensional orientation of given continuous proportion relative to the preceding continuous proportion. Notice that in each case the constraints of equivalently differentiated spatial and temporal perspectives are anti-parallel. The opposite direction of equivalent spatial and temporal constraints causes the geometric forms represented by their respective numeric species to have an inverse relationship.

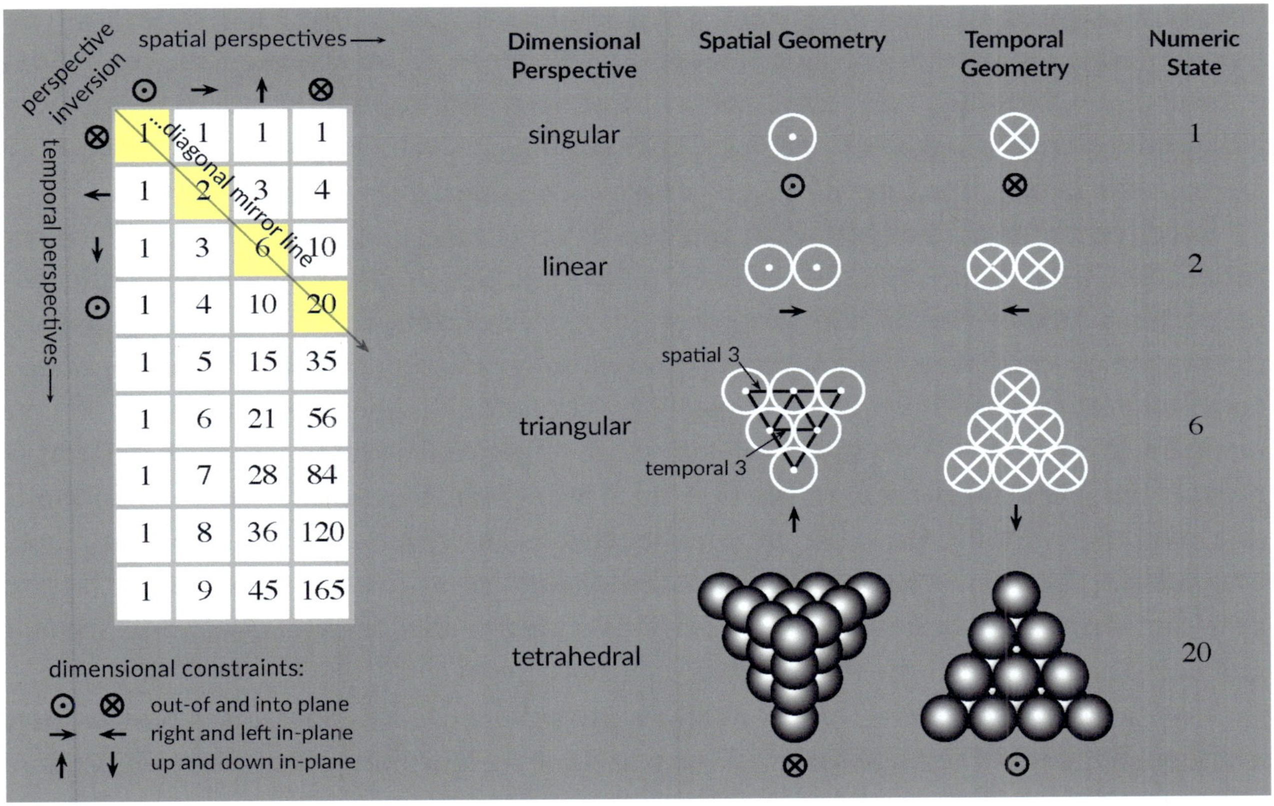

Figure 6. The relative orientation of spatial and temporal geometries represented by the diagonal numeric species of Pascal's Triangle. Dimensional constraint indicators are provided to show the anti-parallel relationship between spatial (column) and temporal (row) continuous proportion geometries (perspectives).

Oppositely constrained spatial and temporal perspectives of the point sum form the first line, the number 2. Since the summation of the two anti-parallel singular perspectives forms the first line whose ends are directionally opposed to one another (180° apart) and this same number 2 participates in the two mirroring continuous arithmetic proportions of Pascal's Triangle, we can infer that the perspectives of space and time directionally oppose on another. But, since the shape of the line is 2-fold, there is no observable difference in geometric orientation between these spatial and temporal continuous proportions to suggest an inverse relationship.

Observation of the triangular number 6, which is composed of the spatial and temporal triangular number 3s, suggests that space and time are reciprocally oriented (inverted) since said triangle is centered about an inner inverted triangle. For instance the spatial triangular number 6 has an inner inverted triangle oriented the same as the temporal triangular number 3 (see Figure 6). The triangular number 6 can consequently morph to either orientation by switching threes between inner and outer triangle. The outer triangle is always oriented to the perspective that has been adopted. This same inner inverted triangle forms the base of the inner inverted tetrahedron at the center of the tetrahedral number 20. Observation of tetrahedral proportions affirms that spatial and temporal perspectives have a negative reciprocal relationship, since their continuous proportions are opposite in directions and their geometries are inverted (*twin tetrahedrons*).

In conventional mathematics sign signifies direction and inversion is expressed by the reciprocal where numerator and divisor are inverted. But in the geometry of continuous proportions neither sign or the reciprocal function are required. The relative geometric orientation of continuous proportions conveys both direction and inversion. Consider a conventional number line (continuous linear proportion) where to the right of zero numbers are positive and to the left numbers are negative. The fact that the two halves of the number line increase in opposite directions is enough to establish their relationship to one another without the need of the negative sign. Now consider two continuous triangular proportions increasing in opposite directions. In this case both direction and inversion expressed because the proportions increase in opposite directions and the geometric orientation of the triangles are opposite as well.

The negative reciprocal mirror-relationship between time and space is no great mystery so it should come as no great surprise that the mirroring continuous proportions of Pascal's Triangle have such a relationship. The relative orientation of space and time can also be observed more directly from one's own experience of them. One can only predict the duration to travel from point A to point B even if the exact distance between them is known in advance – the exact duration can only be determined definitively once it is in the past (looking backward in time). Space is measured from beginning to end (positive or objective), and time is measured from end to beginning (negative or subjective), and the rate of passage of space is measured in units of time (meters per second), and the rate of passage of time in units of space (seconds per meter). And since the natural order of measurement is space first, then time, it seems reasonable to surmise that this order of operation is due to the natural negative reciprocal relationship between space and time.

This order of operation will be applied to the order of dimensional constraint imposition later in this work.

Ultimately I'll show that the negative reciprocal relationship between spatial and temporal continuous proportions is caused by the mirroring void that surrounds the point. And, by applying a governing incremental order to dimensional constraint imposition, I'll show that this same mirroring void is the inversion center between spatial/temporal and temporal/spatial complementary cuboctahedral rotational symmetry transformations that accompany continuous golden proportion radiating from this same central dimensional origin – The energetic *monad* whose dynamic spherical super-symmetry is in harmony with the golden ratio and its negative reciprocal (the golden roots of space and time).

3.3 The Evolution of Dimensional Perspectives

To conceive of how dimensional perspectives evolve consider how a given perspective is formed from its predecessors. A given perspective integrates the former perspective from the vantage that the continued proportion of the former is observable. Take the example of the linear perspective formed by summing the continuous equal proportion of the point: from within the linear perspective the magnitude of a line cannot be seen, for if one looks down the constraint axis of the linear dimension in either spatial or temporal direction all that can be seen is a single point. Next take the example of the triangular perspective: within plane one cannot see a triangle itself as such, rather one sees the continuous proportions of the line and integrates the arithmetic proportions thereof to form the triangular proportion. Next, within the tetrahedral perspective one sees the proportions of the triangular perspective and integrates them to form the tetrahedral proportion. As a general rule then, the form of a given perspective cannot be seen from within said perspective. Instead, the form of a given perspective can be seen only by its orthogonal successor.

So observation of a given continuous proportion requires the orthogonal vantage or perspective caused by the constraint of void on the succeeding dimensional perspectives of the point. Continuous proportions of the type formed by integrating the preceding proportion along a given line of constraint compose a body of perspectives that may be conceived of as increasing in size along a given line of sight. I refer to these continuous proportions as *translational* because they appear to increment in size along a given line, just as with an ordinary number line where growth occurs through translation along an axis. By this definition the continuous proportions of the individual columns and rows of Pascal's Triangle are all translational.

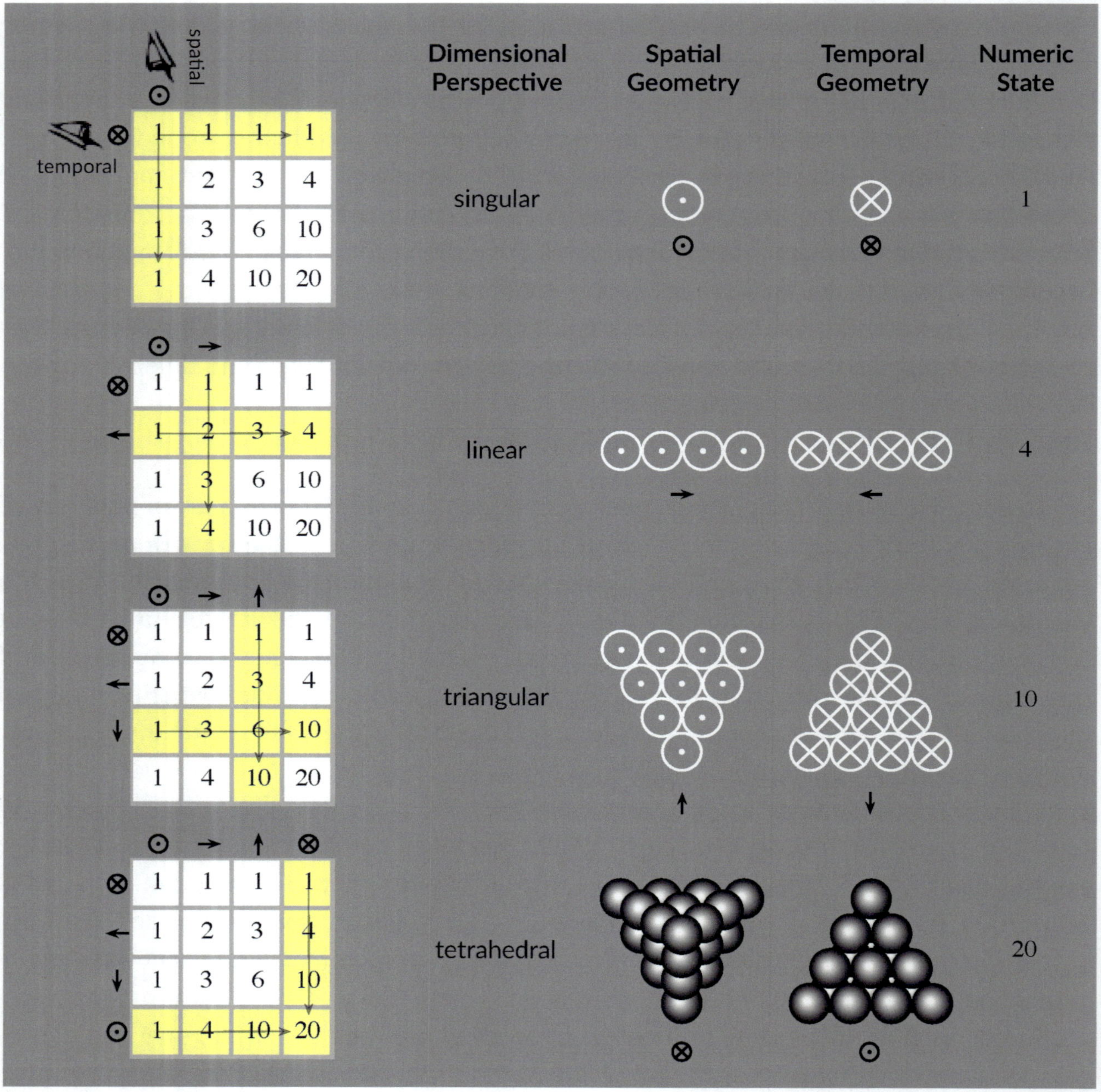

Figure 7. Spatial and temporal translational perspectives of the singular, linear, triangular, and tetrahedral dimensions are highlighted in yellow on the left and their respective geometries are shown on the right. The dimensional constraints that order these geometric perspectives are given as column and row headings.

But what is the perspective of the origin of a given dimension? The formation of this class of perspectives might lead one to presume that the origin of a given translational perspective lies at the beginning of the line of constraint, as in the first point of the line or the vertex of the triangular or tetrahedral perspective, since the integration of the preceding perspective appears to proceed incrementally along said line. Yet the origin of a given translational proportion, the point, is constrained not only by its specific orthogonal orientation relative to the former proportion it observes, but also by the constraints of all preceding perspectives that together with its current constraint define its perspective. Hence, Figure 7 shows that it is the accumulation or compound influence of all constraints that defines a given perspective.

How does the observation of constraint accumulation help us understand the symmetry of the origin of a given translational perspective? First examine the symmetry of singular perspective from the vantage of the linear perspective: from the linear perspective all points reside on the same line of constraint such that only a single point can be seen when looking down the axis of the line. Next examine the linear perspective from the position of the triangular: notice that the point is centered upon the triangular constraint axis together with the centers of every odd number linear form (i.e. the point is at the center of every odd linear number – 1, 3, 5, 7, etc.). Then examine the triangular perspective from the position of the tetrahedral: notice again that the point is centered, this time in a plane of incrementally larger concentric triangles so as to suggest that the point's constraints provide it with triangular intent (i.e. the point is at the center of ever third triangular number – 1, 10, 28, 55, etc.).

These observations affirm that rather than interpreting a translational perspective solely by its present constraint, one must recognize that the origin of a given translational perspective is subject to the compound influence of all integrated constraints that, acting in concert, define its geometric form. So as these cumulative constraints are imposed symmetrically upon the origin, it follows that all perspectives emanate from a central origin – this begins with the initial constraint of inversion between singular spatial and temporal perspectives of said origin. However, even though translational geometries are centered upon a common origin owing to the combined influence of all integrated constraints, the present constraint still defines geometric form orientation and thereby indicates spatial or temporal perspective, which is why proportions appear to be integrated into their successors along a given axis of constraint. Notice that the geometric centers of the species of a given dimension reside on the constraint axis of the next dimension: the centers of all points reside on the constraint axis of the linear dimension, the centers of all lines reside on the constraint axis of the triangular dimension, the centers of all triangles reside on the constraint axis of the tetrahedral dimension, and so on.

As a result of these observations I interpret the point and the line as geometric forms that can have either spatial or temporal orientation just as the two equilateral triangles that compose the "Star of David" motif or as the Platonic twin tetrahedrons that form the star tetrahedron. Yet form orientation is only geometrically apparent beginning with triangular proportions since spatial and temporal points and lines are indistinguishable without any constraint designations. Linear spatial and temporal constraint orientations are evident from the two possible directions along a line and serve as the basis for defining the subsequent triangular and tetrahedral constraint orientations. However, the constraint orientations of singular spatial and temporal translational perspectives can only be inferred by the continuous 6-fold pattern of rotation between the six possible dimensional constraint orientations. The supporting evidence for this will be detailed latter in this work.

Consider the constraint orientations of the singular spatial and temporal translational perspectives in Figure 7. How is it that the point can ever be itself alone, unless it is dimensionally introverted? The spatial and temporal perspectives of the point observe one another across the mirror inversion center that separates them, as shown in Figure 2. And what else can this inversion center between them be but neutral void? This mirror of void constrains the singular spatial perspective to invert to the singular temporal perspective and vice versa.

Further, since the singular spatial continuous equal proportion includes every vantage of temporal translational perspective of the origin and the singular temporal continuous equal proportion includes every vantage of spatial translational perspective of the origin, the perspective of the point has at once a spherical intent, linear intent, triangular intent, tetrahedral intent, and beyond a continuum, affirming ancient hermetic wisdom: *The Oneness then being Source, containeth every number, but is contained by none; engendereth every number, but is engendered by no other one.*

3.4 THE EVOLUTION OF NUMERIC SPECIES

> All things, at least those we know, contain Number; for it is evident that nothing whatever can either be thought or known, without Number. Number has two distinct kinds: the odd, and the even, and a third, derived from the mingling of the other two kinds, the even-odd. Each of its subspecies is susceptible of many numerous varieties, which each manifests individually.
>
> Philolaus* [32]

Here Philolaus makes the important point that there are many numeric species beyond the odd and even of the linear perspective. His observation may be interpreted to mean that the number three, as the combination of one and two, represents a third *even-odd* species when viewed from the triangular versus the linear perspective. The evolution of numeric variety is associated with the degree of separation from the singular perspective

*Philolaus of Croton was educated by Lysis who was a student of Pythagoras, 5th century BCE.

where only one species exists. Beyond even-odd, the reader will have to observe the dimensional geometry of Pascal's Triangle that creates each numeric species (Figure 8) in order to be convinced of the validity of what is proposed here.

Notice that singular translational perspectives indicate that all point symmetries are super-positioned upon the self-same origin. Thus, having continuous equal proportion, the origin requires only one numeric species to describe all point symmetries. As the point differentiates to form a line, two numeric species evolve: odd (1-like) and even (2-like), with every other number only (odd species) having centers that super-position symmetrically upon the singular origin.* Then, as the point differentiates to form the equilateral triangle three numeric species evolve and every third numeric species has a center that coincides with the origin. As the point differentiates further to form the tetrahedron four numeric species evolve and every fourth numeric species has a center that coincides with the origin. The "lines of sight" shown in Figure 8 can be extended continuously since the proportions of Pascal's Triangle are continuous. This implies that the numeric diversity within the continuous proportions that follow the tetrahedral proportion continues to increase and that centered geometries continue to evolve along the highlighted lines of Figure 8.

It follows then that, as the point is the centered origin of every perspective, the numeric species of a given perspective contain centered points that coincide with this centered origin at a frequency determined by the number of dimensional constraints the perspective is subject to. This frequency is indicated by the reciprocal of the first differentiated magnitude (number) of a given continuous proportion: Every singular number is the same central point, every other (1/2) linear number has a central point (1 then 3 then 5...), every third (1/3) triangular number has a central point (1 then 10 then 28...), every fourth (1/4) tetrahedral number (1 then 35 then 165...) and so on. Each successive translational proportion contains a variety of numeric species equal to the number of dimensional constraints imposed upon them: singular has one species (1) and one constraint, linear has two species (odd or 1-like, even or 2-like) and is twice constrained, triangular has three species (1-like, 3-like, and 6-like) and is thrice constrained, and so on. This is shown in the lower left portion of Figure 8 and corresponds with the frequency of the yellow highlighted *centered* geometries.

---

*This is very much like what the Pythagoreans explained with the first principles of *peras* (limited-odd) and *apeiron* (unlimited-even),[33] where the opposing attributes of the monad and the dyad commence the play of opposites – the primordial *isotropic point* and its first offspring the opposing *anisotropic line*. The even dyad is akin to the even mirror symmetry of the infinite tangent void and comes to be after void and point first sum to constrain the arithmetic proportions of spatial and temporal linear perspectives. Hence, the continuous arithmetic proportion has both the finite qualities of the point and infinite qualities of the void.

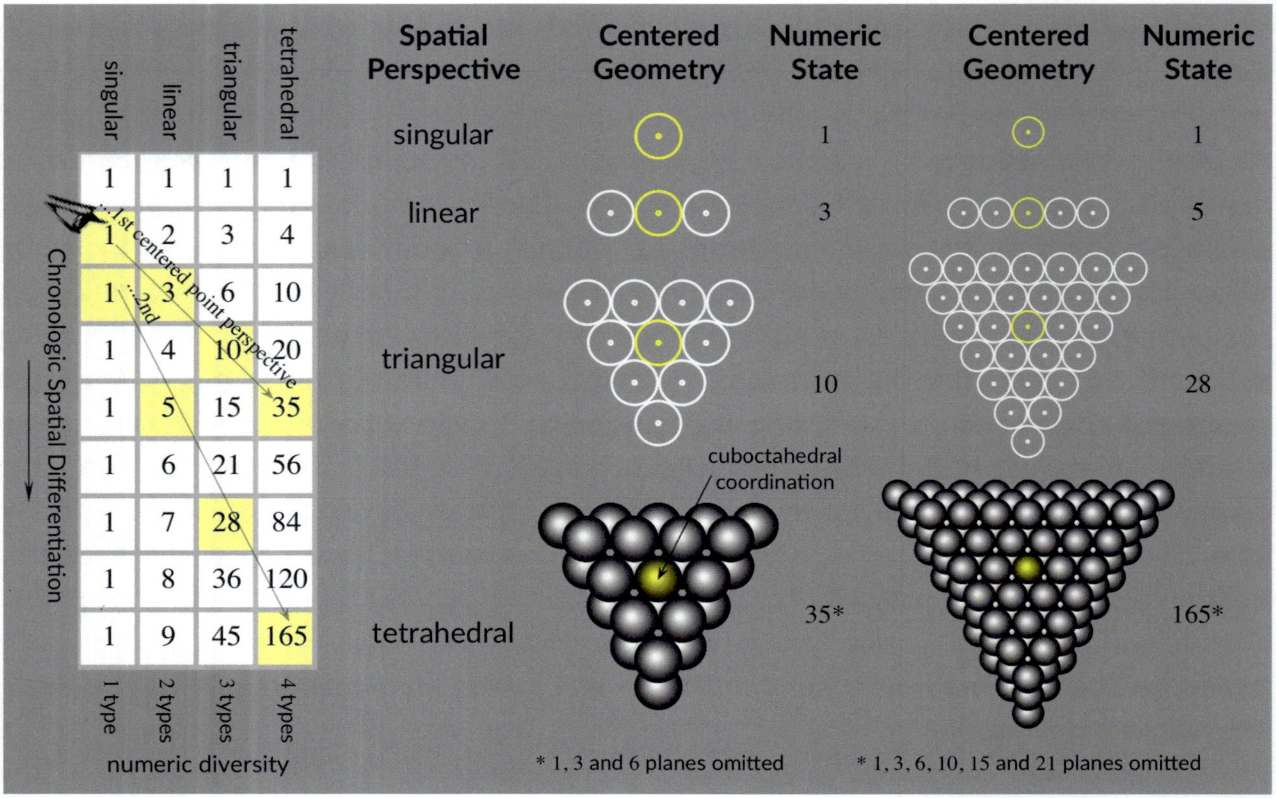

Figure 8. The evolution of the spatial perspectives of the point and the corresponding evolution of numeric diversity: Follow along either "line of sight" of the spatially oriented point observing the dimensional geometries that emanate from the same centered origin. Numeric diversity (types of numbers) contained within each translational proportion is provided in the bottom left corner.

I'm convinced that one of the keys to understanding link between number and geometry lies in the 1-like, 3-like, and 6-like planes of the continuous triangular proportion. The Pythagoreans saw in the triangular number 10 (the *Tetraktys*) the return to the form of the monad, and in so doing; I believe that they perceived the geometric underpinnings of number through the invariance of modulus-9. To me it is compelling that triangular number creates the three precursor planar geometries necessary to subsequently fill volumetric space in the tetrahedral dimension. In crystallography these three planes are viewed in the bulk as having a hexagonal arrangement of close packed spheres with three orientations (often denoted simply as planes A, B, and C)[34], and their significance to modern science cannot be overstated. Crystallographers know this same spherical close-packing structure, so prevalent in the natural world, as the face-centered-cubic lattice structure, which is the crystal structure of gold, platinum, silver, copper, and many other atomic elements. The basic planar "hexagonal" spherical close packing can be seen in Figure 8 where six spheres are arranged about a seventh central sphere shaded yellow so that its geometric dimensional evolution is apparent.

The first tetrahedron whose center is coincident with the point corresponds to the tetrahedral number 35 whose center is the same point that is both the center of the triangular number 10 and the linear number 3. Notice that the tetrahedral perspective of the point it is cuboctahedrally coordinated by twelve surrounding points (six in-plane and three above and below with opposing triangular orientations). When these two opposing triangular orientations are reversed the same central point is temporally oriented versus spatially oriented.

Continuous tetrahedral proportions are the most complex dimensional perspectives observable with spherical close-packing constructs. Beyond tetrahedral, geometric visualization is difficult because spatial and temporal perspectives introvert (fold back in on themselves). Introversion is implied by the continued imposition of orthogonal dimensional constraints. For instance the spatial and temporal constraints of the pentangular dimension are anti-parallel to the respective spatial and temporal constraints of the linear dimension. Therefore the constraints of the pentangular dimension essentially fold back in opposition to the constraints of the linear dimension (anti-parallel). This suggests that there are time-like space, and space-like time dimensions — a topic I'll revisit after I've explained more sophisticated numeric visualization tools. For now, it will suffice to say that dimensional constraints continue to fold and then unfold again owing to a 6-fold rotational constraint pattern that orders the continuous proportions of Pascal's Triangle. This rotating constraint symmetry will be confirmed through geometric analyses in the coming chapters.

CHAPTER CITATIONS

[21] Freeman, Kathleen 1948 *Ancilla to the Pre-Socratic Philosophers* Harvard Univ. Press (USA) p 76

[22] Iamblichus trans. Waterfield, R. 1988 *The Theology of Arithmetic* Phanes Press (USA) p 35

[23] Euclid trans. Heath, L. 2002 *Euclid's Elements* Green Lion Press (USA) p 216

[24] preserved by Photius trans. Guthrie, K.S. 1988 *anonymous biography of Pythagoras - The Pythagorean Sourcebook and Library* Phanes Press (USA) p 138

[25] Triangulo de Pascal en el escrito original de Pascal 2007 Cambridge University Library

[26] Guthrie, K.S. 1988 *The Pythagorean Sourcebook and Library* Phanes Press (USA) p 183

[27] Lao Tzu trans. Mair, V.H. 1990 *Tao Te Ching (Ma-Wang-Tui Manuscripts)* Bantam Books (New York) p 8

[28] Fairbanks, Arthur 1898 *The First Philosophers of Greece* Charles Scribners (New York) pp 134-5

[29] Adair, R. 1987 *The Great Design – Particles, Fields, and Creation* Oxford University Press (Newyork) p 232

[30] ZEUS Collaboration 2008 *Meas. of charged current deep inelastic scat. cross sect. with a longit. polarised electron beam at HERA* http://www-zeus.desy.de/zeus_papers/ZEUS_PAPERS/DESY-08-177.ps p 10

[31] Adair, R. 1987 *The Great Design – Particles, Fields, and Creation* Oxford University Press (Newyork) p 280

[32] Guthrie, K.S. 1988 *The Pythagorean Sourcebook and Library* Phanes Press (USA) p 168

[33] Guthrie, K.S. 1988 *The Pythagorean Sourcebook and Library* Phanes Press (USA) pp 22-23

[34] Kingery et al. 1976 *Introduction to Ceramics*, 2nd Edition, John Wiley & Sons, Inc. (USA) pp 48-52

Beautiful Bonds

> ... it's not possible for two things alone to be beautifully combined apart from some third: some bond must get in the middle and bring them both together. And the most beautiful of these bonds is that which, as much as possible, makes itself and the things bound together one, and proportion is suited by nature to accomplish this most beautifully. For whenever, of three numbers, the middle term of any two of them, whether cubic or square, is such that as the first is to it so is it to the last – and again, conversely, as the last is to the middle so is the middle to the first – then the middle term becomes first and last, while the last and first in turn both become middle terms, so that of necessity it will turn out that all the terms will be the same; and once they've come to be the same in relation to each other, all will be one.
>
> Plato [35]

Pythagorean and subsequent Platonic doctrine underpin much of Euclid's treatment of the subject of continuous proportions. Ten forms of proportions have been established by the *neo-Pythagorean School.*[36] Of those ten I'll focus on arithmetic, geometric, and golden proportions in this chapter. Additionally I'll introduce a "new" generalized form of proportion derived from continuous arithmetic proportion that describes all the continuous translational proportions of Pascal's Triangle.

4.1 Continuous Arithmetic Proportion

> The arithmetic proportion in which the middle term (if we take the minimum of three terms for a proportion) overlaps the first term by a quantity equal to that by which it is itself overlapped by the last term, or (if *a*, *b*, *c*, are the three terms).
>
> $$\frac{c-b}{b-a} = \frac{c}{c} \quad \text{(example: 1, 2, 3)}$$
>
> Matila Ghyka [37]

Here Ghyka follows the Euclidean style of using letters in alphabetic order to distinguish series terms, thus avoiding confusion between the terms and their numeric value. However, as the usage of letters is somewhat cumbersome and abstract, I will use a numeric subscript instead. Given that a *minimum of three terms* is required to define a proportion and given that I am investigating the continuous proportions of spatial-temporal perspectives, it seems reasonable to define the three terms as past $(t-1)$, present (t) and future $(t+1)$ when evaluating the continuous proportions of spatial perspectives defined by temporal increments. The continuous proportions of equivalent temporal perspectives may be expressed by changing subscript designations to spatial increments.

Consider a series of proportioned magnitudes, as many as one pleases, arranged from smallest to largest: $m_1, m_2, m_3, \ldots, m_{t-1}, m_t, m_{t+1}$, where m_{t-1} equals the past magnitude, m_t the present, and m_{t+1} the future. Let the difference of the two larger of three terms be the numerator and the difference of the two smaller of three terms be the denominator of a ratio, *r*, descriptive of the relative proportions of series terms:

$$r = \frac{m_3 - m_2}{m_2 - m_1} \quad \rightarrow \quad r = \frac{m_{t+1} - m_t}{m_t - m_{t-1}} \tag{4.1.1}$$

When the difference between terms is equal, the ratio equals one and the series of numbers is said to have continuous arithmetic proportion, often expressed as a continuous arithmetic proportion equality:

$$m_{t+1} = 2m_t - m_{t-1} \tag{4.1.2}$$

4.2 SPATIAL AND TEMPORAL TRANSLATIONAL PROPORTIONS

Given that the continuous proportions of each and every translational perspective of Pascal's Triangle is constructed from the continuous equal proportions of the point as described by the number one, and given that the first proportion to evolve from the point is the continuous arithmetic proportion of the line, it should be possible to find a general relation descriptive of the interrelated proportions of each perspective. Furthermore, this relation or equation should be descriptive of continuous arithmetic proportion, and should also contain terms relating the inter-dimensional dependence of each dimensional perspective so that it will be inclusive of all translational perspectives. In order to add this inter-dimensional aspect and to distinguish both a given perspective and its degree of differentiation, I will use a temporal subscript for the chronologic differentiation of a given perspective, and said perspective will be defined by a superscript term that I'll keep separate by placing it before the magnitude, *m*, as follows:

$${}^{perspective}m_{differentiation}, \text{ where a given spatial perspective is ... } {}^{space}m_{time} \text{ or } {}^{s}m_{t}$$

Based on this new notation the linear spatial perspective (*s*=2) is written in terms of the continuous arithmetic proportion equality as follows:

$${}^{2}m_{t+1} = 2\left({}^{2}m_{t}\right) - {}^{2}m_{t-1}$$

However, the preceding equation is still strictly linear. It is lacking a term that links the proportions of a given dimension to preceding proportions such that the relation magnifies proportionate terms appropriately. Observe that arithmetic proportion is inherently linked to the singular dimension because the difference between terms is equal to one. But what makes the difference between terms remain constant is the relationship between the singular perspective and the perspective of void that precedes it.

A term relating the proportion of the linear dimension to that of the void can be added without changing the net relationship between sequential terms of the linear dimension since the void is continuously equal to zero. In the case of the triangular proportion the difference between terms increases incrementally by one since summing the arithmetic proportion that preceded it forms the triangular proportion (i.e. 0, 1, 3, 6, 10, ... has differences between terms of 1, 2, 3, 4, ...). Hence, the singular proportion in effect magnifies the difference between terms to form the triangular proportion since the difference between terms increases incrementally by one. This incremental increase can be accounted for by adding a component to the arithmetic proportion equality relating a given term to the term of the perspective that precedes the prior (s-2). Therefore, the continuous proportions of spatial translational perspectives may be generalized as follows:

$${}^{s}m_{t+1} = 2 \cdot {}^{s}m_{t} - {}^{s}m_{t-1} + {}^{s-2}m_{t+1} \tag{4.2.1}$$

Boundary conditions: Dimensions are bounded by void so in general for null chronologic differentiation the *s*-2 term is also null. Although continuous proportions can be extended to negative terms with this equation, this is the same as switching to temporal perspectives. So in general this equation is descriptive of continuous spatial proportions bounded by void. Temporal perspectives are similarly described in terms of past, present, and future spatial increments with a boundary condition of void.

Figure 9 is provided as proof that the *continuous space-time proportion equality* (equation 4.2.1) describes the interconnected continuous proportions of all the translational perspectives of Pascal's Triangle. Wherein the terms of a given proportion are incremented by an inter-dimensional term from the perspective of the spatial dimension that precedes the former perspective (*s*-2) thereby linking even to even perspectives and odd to odd.* Specifically Figure 9 shows that the triangular spatial perspective (three directional) is magnified by the singular spatial perspective (one directional), and that the tetrahedral spatial perspective (four directional) is magnified by the linear spatial perspective (two directional), and so on.

Three Directional Spatial Proportion
(triangular spatial perspective: s=3)

three directional				three directional		three directional		one directional
$^{s}m_{t+1}$	=	2	⊗	$^{s}m_{t}$	⊖	$^{s}m_{t-1}$	⊕	$^{s-2}m_{t+1}$
1	=	2	⊗	0	⊖	0	⊕	1
3	=	2	⊗	1	⊖	0	⊕	1
6	=	2	⊗	3	⊖	1	⊕	1
10	=	2	⊗	6	⊖	3	⊕	1
15	=	2	⊗	10	⊖	6	⊕	1
21	=	2	⊗	15	⊖	10	⊕	1
28	=	2	⊗	21	⊖	15	⊕	1
36	=	2	⊗	28	⊖	21	⊕	1
45	=	2	⊗	36	⊖	28	⊕	1

Four Directional Spatial Proportion
(tetrhedral spatial perpective: s=4)

four directional				four directional		four directional		two directional
$^{s}m_{t+1}$	=	2	⊗	$^{s}m_{t}$	⊖	$^{s}m_{t-1}$	⊕	$^{s-2}m_{t+1}$
1	=	2	⊗	0	⊖	0	⊕	1
4	=	2	⊗	1	⊖	0	⊕	2
10	=	2	⊗	4	⊖	1	⊕	3
20	=	2	⊗	10	⊖	4	⊕	4
35	=	2	⊗	20	⊖	10	⊕	5
56	=	2	⊗	35	⊖	20	⊕	6
84	=	2	⊗	56	⊖	35	⊕	7
120	=	2	⊗	84	⊖	56	⊕	8
165	=	2	⊗	120	⊖	84	⊕	9

Five Directional Spatial Proportion
(penta-translational spatial perspective: s=5)

five directional				five directional		five directional		three directional
$^{s}m_{t+1}$	=	2	⊗	$^{s}m_{t}$	⊖	$^{s}m_{t-1}$	⊕	$^{s-2}m_{t+1}$
1	=	2	⊗	0	⊖	0	⊕	1
5	=	2	⊗	1	⊖	0	⊕	3
15	=	2	⊗	5	⊖	1	⊕	6
35	=	2	⊗	15	⊖	5	⊕	10
70	=	2	⊗	35	⊖	15	⊕	15
126	=	2	⊗	70	⊖	35	⊕	21
210	=	2	⊗	126	⊖	70	⊕	28
330	=	2	⊗	210	⊖	126	⊕	36
495	=	2	⊗	330	⊖	210	⊕	45

Six Directional Spatial Proportion
(sexa-translational spatial perspective: s=6)

six directional				six directional		six directional		four directional
$^{s}m_{t+1}$	=	2	⊗	$^{s}m_{t}$	⊖	$^{s}m_{t-1}$	⊕	$^{s-2}m_{t+1}$
1	=	2	⊗	0	⊖	0	⊕	1
6	=	2	⊗	1	⊖	0	⊕	4
21	=	2	⊗	6	⊖	1	⊕	10
56	=	2	⊗	21	⊖	6	⊕	20
126	=	2	⊗	56	⊖	21	⊕	35
252	=	2	⊗	126	⊖	56	⊕	56
462	=	2	⊗	252	⊖	126	⊕	84
792	=	2	⊗	462	⊖	252	⊕	120
1287	=	2	⊗	792	⊖	462	⊕	165

Figure 9. Dimensional number proof of the continuous space-time proportion equality: The figure shows that when an inter-dimensional term (perspective *s*-2) is added to the continuous arithmetic proportion equality, the relation is descriptive of all the translational perspectives of Pascal's Triangle.

*The linking of odd and even continuous proportions meshes with the qualities of *peras* (limit-odd) and *apeiron* (unlimited-even) espoused by the ancient Greeks.

Based on the preceding analysis of translational proportions one could make a strong argument that dimensions begin with void rather than point, but it seems more accurate to say that point and void together perpetuate the differentiation of perspective. They are related to one another as that which is differentiated (the point) is related to that which causes differentiation (void). In other words point is to void as differentiated is to differentiator.

4.3 CONTINUOUS GEOMETRIC PROPORTION

The topic of continuous geometric proportion lays the foundation for the topic of integral dimensionality and complementary super-symmetry associated with the rotation of dimensional constraints, versus those of translational perspectives associated with continuous arithmetic proportion. Like translational perspectives, the integral perspective of continuous geometric proportion begins with the point and can be evaluated using Pascal's Triangle. But I still have some ground to cover before I present these concepts in detail. For now I will simply build the mathematical form and cement it in context later. Again from Ghyka:

> ...the equality *a/b=c/d* ...is the *geometrical proportion*, called discontinuous in the general case when *a*, *b*, *c*, *d*, are different, and continuous geometrical proportion if two of these numbers are identical. The typical continuous (geometrical) proportion is therefore
>
> $$\frac{a}{b}=\frac{b}{c} \quad \text{or} \quad b^2=ac \quad \text{...(also given as)} \quad \frac{c-b}{b-a}=\frac{c}{b} \quad \text{(example: 1, 2, 4)}$$
>
> Matila Ghyka [38]

Recall that in the case of arithmetic proportions the difference between terms is equal. So the ratio descriptive of said proportion is unity ($r=1$). With continuous geometric proportion the descriptive ratio is equal to the ratio of any two consecutive terms. And, since I defined the ratio to be the difference between the largest and middle term divided by the difference between the middle and smallest term, the descriptive ratio becomes larger over smaller consecutive terms. Using the previously established mathematical notation yields the following continuous geometric proportion and equality:

$$\frac{m_{t+1}}{m_t}=\frac{m_{t+1}-m_t}{m_t-m_{t-1}}=r \tag{4.3.1}$$

$$m_t^{\,2}=m_{t-1}m_{t+1} \tag{4.3.2}$$

Continuous geometric proportion, commonly referred to as exponential growth, begins with the continuous doubling of the number one:

$$\frac{2^1}{2^0}=\frac{2^2}{2^1}=\frac{2^3}{2^2}=\frac{2^4}{2^3}=\frac{2^5}{2^4}=\frac{2^6}{2^5}\cdots\frac{2^{n+1}}{2^n},\text{ or }\frac{2}{1}=\frac{4}{2}=\frac{8}{4}=\frac{16}{8}=\frac{32}{16}=\frac{64}{32}\ldots\text{ and so on,}$$

where n may be considered a certain interval that distinguishes series terms.

The preceding example is an exponential expansion of base-2, but any number raised to a series of consecutive powers will be in continuous geometric proportion. The next is base-3:

$$\frac{3^1}{3^0}=\frac{3^2}{3^1}=\frac{3^3}{3^2}=\frac{3^4}{3^3}=\frac{3^5}{3^4}=\frac{3^6}{3^5}\cdots\frac{3^{n+1}}{3^n},\text{ or }\frac{3}{1}=\frac{9}{3}=\frac{27}{9}=\frac{81}{27}=\frac{243}{81}=\frac{729}{243}\ldots\text{ and so on}$$

Every form of continuous geometric proportion begins with the number one since any number raised to the null power aside from zero is equal to one by convention. Also, the base of the continuous geometric proportion is always equal to its descriptive ratio.

4.4 Continuous Golden Proportion

> There are in all ten terms of proportions, established by the neo-Pythagorean School. The tenth corresponds to the additive series 1, 1, 2, 3, 5, 8, 13, 21, 34, 55, ... et cetera, in which each term is equal to the sum of the two preceding ones; it is intimately connected with the Golden Section and plays an eminent role in Botany.
>
> $$\frac{c-a}{c-b}=\frac{b}{a}\quad\text{(example: 3, 5, 8)...Fibonacci series}$$
>
> Matila Ghyka [39]

Recall that what led to my exploration of Pascal's Triangle is its relationship with the Golden Ratio (*Golden Section*) and golden proportions such as the Fibonacci series, since if humans are microcosms and life is more than a statistical probability, then the symmetry of life must logically be contained within the origin of All. And, as Ghyka states in the preceding quote, the Golden Ratio and golden proportion *play an eminent role in Botany*.

Subsequent analysis will show that, like continuous geometric proportion, continuous golden proportion is associated with a certain vantage of the point where the rotating dimensional constraints of Pascal's Triangle form super-symmetric spatial/temporal intervals that can be interpreted geometrically. As with continuous geometric proportion, I am expounding the mathematics of continuous golden proportion in advance of a thorough analysis of this geometric context.

Golden proportion is the special case of geometric proportion where, of the three terms required to define a proportion, the third term is the sum of the preceding two terms. The ratios that describe this form of proportion are the Golden Ratio and its negative reciprocal, and they are associated with the ratio of two consecutive terms, the larger being the numerator and the smaller the denominator of the ratio. So to define continuous golden proportion one begins with continuous geometric proportion (equation 4.3.1) and substitutes the sum of the first and second terms for the third as follows:

1) Letting $m_{t+1} = m_t + m_{t-1}$ gives continuous golden proportion.

$$\frac{m_t + m_{t-1}}{m_t} = \frac{m_{t-1}}{m_t - m_{t-1}} \tag{4.4.1}$$

2) Rearranging yields a polynomial equation with respect to the ratio of two consecutive terms.

$$\left(\frac{m_t}{m_{t-1}}\right)^2 = \frac{m_t}{m_{t-1}} + 1 \tag{4.4.2}$$

3) Assigning $\frac{m_t}{m_{t-1}}$ as the characteristic ratio (*r*) highlights the polynomial equation.

$$r^2 = r + 1 \tag{4.4.3}$$

4) Solving for *r* using the Quadratic Formula gives two solutions, one positive and the other its negative reciprocal, two golden roots of the polynomial equation.

$$r = \frac{1 \pm \sqrt{5}}{2} \quad \left\{ \begin{array}{ll} \varphi_{objective} : \ \varphi_+ = \dfrac{1+\sqrt{5}}{2} & (4.4.4) \\[2ex] \varphi_{subjective} : \ \varphi_- = \dfrac{1-\sqrt{5}}{2} & (4.4.5) \end{array} \right.$$

Continuous golden proportions arise from the melding of the golden roots either by the difference of their powers or by their sum, where their power (*n*) may be considered a certain interval that defines the terms of the given golden exponential series:

$$\text{Fibonacci Series} = \frac{\varphi_+{}^n - \varphi_-{}^n}{\varphi_+ - \varphi_-} = \frac{\varphi_+{}^n - \varphi_-{}^n}{\sqrt{5}} \tag{4.4.6}$$

$$\text{Lucas Series} = \frac{\varphi_+{}^n + \varphi_-{}^n}{\varphi_+ + \varphi_-} = \frac{\varphi_+{}^n + \varphi_-{}^n}{1} \tag{4.4.7}$$

Generally speaking the negative root of quadratic equations are treated as imaginary, being a result of the paradox that a negative multiplied by a negative is a positive; this is akin to ignoring the temporal perspective. Both sides of the coin together make the whole. The derivation of the golden ratio produces a *thesis*, the Golden Ratio, and an *antithesis*, its negative reciprocal. The fact that these two roots sum to one suggests that, as the singular origin of space-time, the point has the capacity to harbor these two opposing symmetry elements. Given their negative reciprocal relationship, which I've associated with the relationship between spatial and temporal perspectives, it follows that the positive root is an objective spatial symmetry element, and that the negative root is a subjective temporal symmetry element.*

Often in mathematical treatments of this subject I find that the negative reciprocal of the golden ratio is set aside and continuous golden proportions are described using just the positive root since it can be manipulated into its negative reciprocal. For instance Binet's Fibonacci Number Formula[40] is like equation 4.4.6, but uses only the positive solution to the Golden Ratio. Per equations 4.4.6 and 4.4.7, the two roots together are required to describe continuous golden proportion. This provides an explanation for why the ratio of consecutive terms of the Fibonacci series only approximates the positive root rather than being exact, as would be the case if one ratio alone were the descriptor of proportion such as the ratio of consecutive terms of continuous geometric proportion. Continuous geometric proportion is defined by one exponential base (one descriptive ratio), whereas continuous golden proportions are defined by the merging of two exponential bases (two descriptive ratios), a base and an anti-base (*golden roots*). The merging of base and anti-base form continuous golden proportions in two ways: when the

---

*The *golden roots* of space and time are analogous to the ancient Chinese notions of the primordial symmetry elements *yang* and *yin*.

difference between equivalent powers of base and anti-base divided by the difference between base and anti-base are arranged in the arithmetic order of powers (Fibonacci series), and when the sum of equivalent powers of base and anti-base divided by the sum of base and anti-base are arranged in the arithmetic order of powers (Lucas series).

Since continuous geometric proportion is defined by one base and continuous golden proportion is defined by two bases, and since there is a negative reciprocal relationship between these two bases, it is possible to anticipate that the latter is likely the result of a combined spatial/temporal perspective and the former is not. In other words, one may expect continuous geometric proportion to be formed from rotation intervals defined by either spatial or temporal dimensional constraints, and continuous golden proportion to be formed from rotation intervals defined by a combination of both spatial and temporal dimensional constraints.

CHAPTER CITATIONS

[35] Plato *Timaeus* trans. Kalkavage, Peter 2001 Focus Publishing (USA) p 61-62
[36] Ghyka, Matila 1977 *The Geometry of Art and Life* Dover Publications (New York) p 4
[37] Ghyka, Matila 1977 *The Geometry of Art and Life* Dover Publications (New York) p 4
[38] Ghyka, Matila 1977 *The Geometry of Art and Life* Dover Publications (New York) p 4
[39] Ghyka, Matila 1977 *The Geometry of Art and Life* Dover Publications (New York) p 5
[40] Weisstein, Eric W. "Binet's Fibonacci Number Formula." From *MathWorld* (web resource)

Elementary Numbers

> And the Pythagoreans say that there is void, and that it enters into the heaven itself from the infinite air, as though it (the heaven) were breathing; and this void defines the nature of things, inasmuch as it is a certain separation and definition of things that lie together; and this is true first in the case of numbers, for the void defines the nature of these.
>
> Aristotle [41]

> That number admits nothing beyond the ennead (nine), but rather everything circles around within it, is clear from the so-called recurrences: there is natural progression up to it, but after it there is repetition. …so that it is by no means possible for there to subsist any number beyond nine elementary numbers.
>
> Iamblichus [42]

Per Aristotle, the Pythagoreans held that the void defines the nature of numbers; this corresponds with my previous position that the sums of void and point create the orthogonal translational continuous proportions of Pascal's Triangle. Following Pythagorean doctrine, Iamblichus highlights the invariant nature of the number nine and the analog *recurrences* of modulus-10 numbers evident when they are reduced* to nine residual *elementary* modulus-9 numeric species. In modulus-9 (a.k.a. base-9) the void (zero) is the first multiple of nine and hence is equivalent to nine from a modulus-9 symmetry perspective. So the statements of Aristotle and Iamblichus are consistent with one another as cause is to effect since *void defines the nature of* number through its causal geometric link to the continuous proportions of Pascal's Triangle with the effect being the analog *recurrences* of the *nine elementary numbers* within these proportions. In other words, since the orthogonal perspectives of void to point are responsible for the numeric proportions of the translational perspectives of Pascal's Triangle, it is evident that the number zero causes the numerically symmetrical differentiation of number from the number one, thereby defining the nature of number.

---

*Numeric reduction is the process of summing the digits of a number until only one digit remains. It is detailed in section 5.2.

I will show that continuous whole number proportions are symmetric with respect to modulus-9 since the orthogonal summation of void and point causes the perspective of the point to be constrained or held invariant to the perspective of void. Hence, the orthogonal juxtaposition of void to point causes the numeric invariance associated with modulus-9. The concept of invariance is familiar to those who study symmetry, since the observation of symmetry requires that something be held invariant such as mirror plane, axis of rotation, center of inversion, line of sight, etc., so that relationships may be established with respect to a fixed geometric feature. Invariance, then, defines the perspective wherefrom the symmetry is evident.

I will show that, in the context of dimensional number, modulus-9 is the vantage from which numeric symmetry is apparent and that the number 9 is the invariant numeric species integrated within translational perspectives when void and point are summed. In my efforts to establish a unifying axiomatic perspective wherefrom the causal link between dimensional geometry and number symmetry is self-evident I have found modulus-9 mathematics to be the single most informative analysis tool at my disposal.

5.1 Tetraktys: The Perennial Fount and Root of Nature

The governing role of modulus-9 numeric symmetry is readily apparent through inspection of the triangular number series. The influence of modulus-9 is most evident with the triangular number 10 (the *Tetraktys*), which marks the first analog *recurrence* of the number 1 within a repeating pattern of three distinct types of triangular planes. So it is no wonder the Pythagoreans considered it the source of all their wisdom and the *root of Nature.*[*] [43]

The triangular number ten is composed of nine points arranged about one. It is the second A-layer of the A-B-C spherical close-packed stacking sequence that forms the cuboctahedral space-filling geometry discussed previously and shown in Figure 10: both the numbers 10 and 1 represent layer-A symmetry (10 can be reduced to 1 symmetrically). Similarly, layer-7 (28) can be reduced to layer-4 and then to layer-1.

Each consecutive plane of a given type is formed by addition of a concentric triangle about the previous layer of the same type. Every third triangular plane beginning with the initial 1 belongs to the same 1-like species of layer-A number. The layer-A triangles are consecutive sums of the series 1, 9, 18, 27, and so on (1, 10, 28, 55, ...). Layer-B triangles form the 3-like species of number which are consecutive sums of the series 3, 12, 21, 30,

[*]There are clear references to the importance of both modulus-9 and the triangular number 10 in ancient literature, but it seems the link between the two was either never fully realized or perhaps more likely was lost due to the exclusive oral tradition of the Pythagorean mystery school.

and so on (3, 15, 36, 66, ...). Likewise layer-C triangles form the 6-like species which are sums of the series 6, 15, 24, 33, and so on (6, 21, 45, 78, ...). All A-triangles reduce to 1 since they are sums of 1 and multiples of 9. And when reduced, B-triangles alternate 3-6-9 as sums of a series of numbers that each reduce to three, and C-triangles alternate 6-3-9 as sums of a series of numbers that each reduce to six. Therefore triangular number symmetry affirms the existence of a fundamental physical connection between spherical close-packing geometry, continuous proportions, and modulus-9 number symmetry. Remember that all the continuous proportions of Pascal's Triangle are intimately connected so the readily apparent modulus-9 symmetry of the continuous triangular proportion means that all the continuous proportions of Pascal's Triangle are governed by modulus-9 symmetry transformations.

Modulus-10 Triangular Number Series: 1, 3, 6, 10, 15, 21, 28, 36, 45, 55, 66, 78, 91, 105, ...
- A-Planes: 1, 10, 28, 55, 91, 136, ...
- B-Planes: 3, 15, 36, 66, 105, 153, ...
- C-Planes: 6, 21, 45, 78, 120, 171, ...

Modulus-9 Triangular Number Series: 1, 3, 6, 1, 6, 3, 1, 9, 9, 1, 3, 6, 1, 6, 3, 1, 9, 9, ...
- A-Planes: 1, 1, 1, 1, 1, 1, ...
- B-Planes: 3, 6, 9, 3, 6, 9, ...
- C-Planes: 6, 3, 9, 6, 3, 9, ...

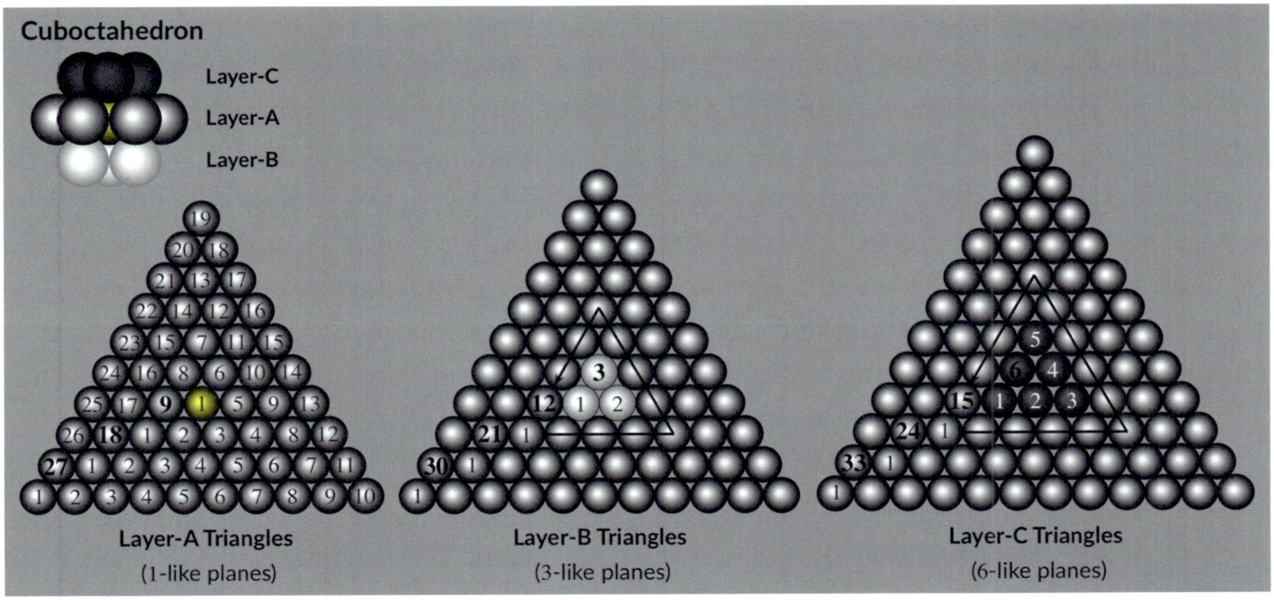

Figure 10. Triangular number modulus-9 relationship: A-planes begin with 1 and increment by 9n (1+9x0 = 1, 1+9x1=10, 10+9x2=28, 28+9x3=55). B-planes begin with 3 and increment by 9n+3. C-planes begin with 6 and increment by 9n+6. In this context n represents the nth triangle of a given species beginning with 0.

5.2 CASTING OUT NINES

> ...10 becomes a monad by subtraction of one elementary quantity (i.e. one ennead), and again 11 and 20 become a dyad by subtraction of either one or two enneads, and 12 and 30 become a triad, and again 100 becomes a monad, when 11 enneads are subtracted, and so on *ad infinitum*...
>
> Iamblichus [44]

The method of *casting out nines* (numeric reduction) is of ancient origin. Today most people associate numeric reduction with numerology, where all of the digits of a given number are summed until only one digit remains. Yet prior to the age when the positional numeral system (base-10 or modulus-10) was developed (c. 600-800 CE), Brahmin mathematicians used only the natural numbers 1 through 9.[45] After the positional numeral system was developed, the method of casting out nines was commonly used for checking arithmetic performed on an abacus.[46] Since modulus-9 preceded modulus-10, it is likely that modulus-9 remained in use after modulus-10 was adopted, as a partner to it owing to its unique properties.

The term *positional* refers to the 1s, 10s, 100s, ... positions occupied by digits within a given number that when summed equal said number. For instance the number 497 is $400+90+7$. Whereas the same number in modulus-9 only has one position that conveys its proximity to a multiple of nine. For instance, the number 497 becomes 2 in modulus-9 since the closest 9-multiple is 495 (9 times 55) and $497-495=2$. So in this example 55 nines are *cast-out*. The short cut to this solution is numeric reduction, where $4+9+7=20$ followed by $2+0=2$.

C.F. Gauss classified the arithmetic properties of modulus-9 within the greater context of his congruent number theory.[47] This branch of mathematics characterizes numeric behavior with respect to *moduli* (divisors) that generate predictable patterns of *residues* (remainders), which are further grouped into *residue classes*. Any multiple of nine has no residue in a modulus-9 context, and therefore *reduces* to nine (or by convention zero). All numbers that are not nine-multiples have remainders with respect to a divisor of nine and their possible residues are 1 thru 8. By extension, any arithmetic operation involving only nine-multiples yields a quantity that has a null remainder with respect to a divisor of nine. Consequently, I contend that the number nine has the property of invariance.

There are five complementary pairs of decadic numbers: 0-9, 1-8, 2-7, 3-6, and 4-5 (each pair sums to the invariant nine). With modulus-9, these reduce to four pairs (1-8, 2-7, 3-6, and 4-5) plus the neutral sum of any pair of complements, 9. Since all complementary modulus-9 residues sum to nine not zero I will depart from standard number theory texts in this work and use nine in place of zero to represent cases where there is no residue with respect to divisor of nine. Also, a negative modulus-9 residue may be converted to its positive complement as follows: $-2+9=7$. This concept is like the concept of absolute value, though it can be more accurately described as absolute residue. As an extension of absolute value, I'll use double vertical bar-brackets to indicate the dual mathematical operation of casting out nines and negative sign conversion. The following number-line example illustrates both using 9 in place of 0 and complement sign conversion:

MODULUS-10 ...

$$\ldots,-12,-11,-10,-9,-8,-7,-6,-5,-4,-3,-2,-1,\mathbf{0},1,2,3,4,5,6,7,8,9,10,11,12,\ldots$$

MODULUS-9 WITH NEGATIVES ...

$$\ldots,-3,-2,-1,-9,-8,-7,-6,-5,-4,-3,-2,-1,\mathbf{9},1,2,3,4,5,6,7,8,9,1,2,3,\ldots$$

MODULUS-9 WITHOUT NEGATIVES ...

$$\ldots,6,7,8,9,1,2,3,4,5,6,7,8,\mathbf{9},1,2,3,4,5,6,7,8,9,1,2,3,\ldots$$

Notice that with modulus-9, the four pairs of complements are equidistant from neutral centers when arranged arithmetically as follows: 5-6-7-8-**9**-1-2-3-4. This highlights that the modulus-9 residues 5 thru 8 have a subtractive (negative or subjective) quality and the modulus-9 residues 1 thru 4 have an additive (positive or objective) quality. For example: when objective residue-1 is added to any number the resulting modulus-9 residue is incremented by 1 ($\|1+20\| \Rightarrow 3$), whereas when the subjective residue-8 is added to any number the resulting modulus-9 residue is decremented by 1 ($\|8+20\| \Rightarrow 1$), unless of course either is added to null residue-9 in which case each retains its own original residue. Alternatively this behavior can be conceived of precession and recession in the context of an analog number wheel, since residues 1 thru 4 cause precessions and residues 5 thru 8 cause recessions in the residue of that with which they are summed:

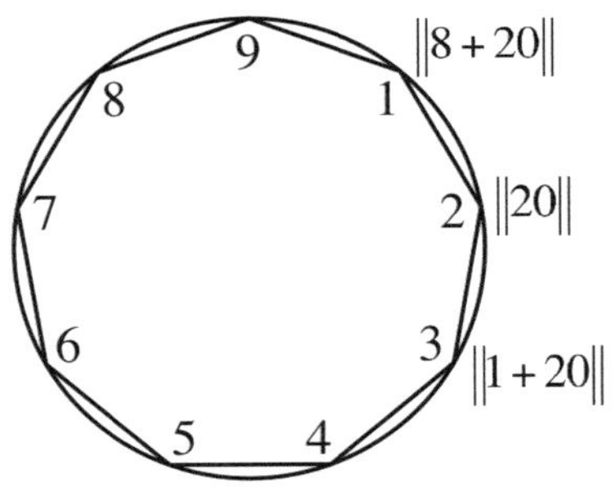

Perhaps the most familiar form of modulus-9 complementary symmetry is its use in geometry, where a circle is divided into 360°. In this form complementary angles such as 10° and 80° reduce to the modulus-9 complements 1 and 8.

The nine modulus-9 residues are composed of three classes of modulus-3: 1-4-7, 2-5-8, and 3-6-9.[48] I designate these residue classes to be primary, secondary, and tertiary phases respectively.[*] I have equated the primary modulus-9 number phase 1-4-7 with the colors yellow-blue-red, the secondary number phase 2-5-8 with green-orange-violet, and the tertiary number phase 3-6-9 with white-black-grey. Next I will show how complementary modulus-9 numeric behavior is analogous to color symmetry, why I consider each of the nine modulus-9 numbers to be color-number polarities, and why each complementary pair can be considered a dipole in a cuboctahedral context. So from this point forward I will use the term *polarity* instead of residue when referring to the *nine elementary numbers* in the modulus-9 context.

5.3 Modulus-9 Complementary-cuboctahedral Color-Number Dipoles

> For unity in the details harmonizes all the parts of a whole, as by the participation of the First Cause.
>
> Porphyry [†] [49]

The notion that the energy singularity harbors fundamental constituents required to create the order of the cosmos suggests that the logic of modulus-9 must somehow be an aspect of *first cause*, a super-symmetry of the singular state perhaps. And further, it suggests that this symmetry governs every aspect of manifestation in perpetuity. Super-symmetry is analogous to what many scientists refer to as vacuum symmetry or the symmetry of empty space, and yet it differs in the sense that at present I am evaluating the perspective of energy and its symmetry versus the perspective of matter and its symmetry. At present the scientific community doesn't distinguish between the two and is focused on trying to understand natural phenomena experimentally from a matter or integral energy/anti-energy perspective. I would suggest, however, that an awareness of the dimensional context (perspective) of one's experimental findings is vital to accurately interpreting them.

Previously I showed how the centered origin of spatial and temporal perspectives has an underlying cuboctahedral geometry composed of 12 spheres about 1. Following this line of reasoning my presumption is that the natural geometric coordination of the point is associated with its governing super-symmetric perspective. The governing super-symmetric perspective of the point is one that integrates all dimensional perspectives within a single continuous proportion geometry that controls how dimensional

---

[*]The terms primary and secondary have been adopted from accepted color theory to which tertiary is added to describe the white, black, and grey color-number phase.

[†]Porphyry was Neo-platonic philosopher in the 3rd century CE who taught Iamblichus.

perspectives differentiate or unfold. In general a super-symmetric perspective is one that constrains perspectives of lesser or subservient symmetries. The *governing* super-symmetric perspective has the highest level of symmetry, and it constrains all other perspectives to its symmetry. In the coming chapters I'll show that super-symmetric perspectives are constrained by rotational symmetry transformations about a given axis of symmetry. This same complementary geometric symmetry can be derived from both triangular number symmetry and from published color symmetry.

Triangular number symmetry highlights the complementary cuboctahedral coordination of the point since the point is centered between 3-like and 6-like triangular planes that form opposing cuboctahedral faces (a 3-6 dipole). Given this geometry, the relative geometric orientation of the other numeric complements can be determined since the reduced sum of the primary modulus-9 phase has polarity-3: $\|1+4+7\|$, and the reduced sum of the secondary modulus-9 phase has polarity-6: $\|2+5+8\|$, which suggests that in both cases the three polarities are adjacent cubic faces that share a common vertex equal to their sums (octahedral faces of polarity 3 and 6).

This observation is consistent with Itten's color symmetry model,[50] although he depicts color symmetry as planar hexagonal. Nevertheless his color symmetry conforms to complementary-cubic geometry with respect to a 6-fold cubic axis of symmetry (cube diagonal) when the confluence of primary phase cubic faces is designated as the back vertex, and the confluence of secondary phase cubic faces is designated as the front vertex (or vice versa) -- this interpretation places color complements on opposing cube faces (see Figure 11). So, when the primary number phase is equated with the primary color phase and the secondary number phase is equated with the secondary color phase, the cubic vertex of the primary color-number phase corresponds to the polarity-3 octahedral face and the color white, and the cubic vertex of the secondary color-number phase corresponds to the polarity-6 octahedral face and the color black. This leaves nine to be designated as neutral grey, as are the sums of color complements.[51]

In Figure 11 polarity-1 is designated yellow, since clockwise counting with respect to the primary phase vertex begins at the 12 o'clock position and since this is where Itten places yellow on what will henceforth be referred to as the *color cube*. Notice that equating polarity-1 with yellow is a somewhat arbitrary selection based on convenience. Any primary phase color could have been selected with equivalent results. The remaining colors are incrementally assigned in conformance with modulus-3 residue class phases discussed previously. The primary phase is designated as the back vertex and rotated to the front in Figure 11. This switches the position of blue and red relative to the color cube at the top of the figure.

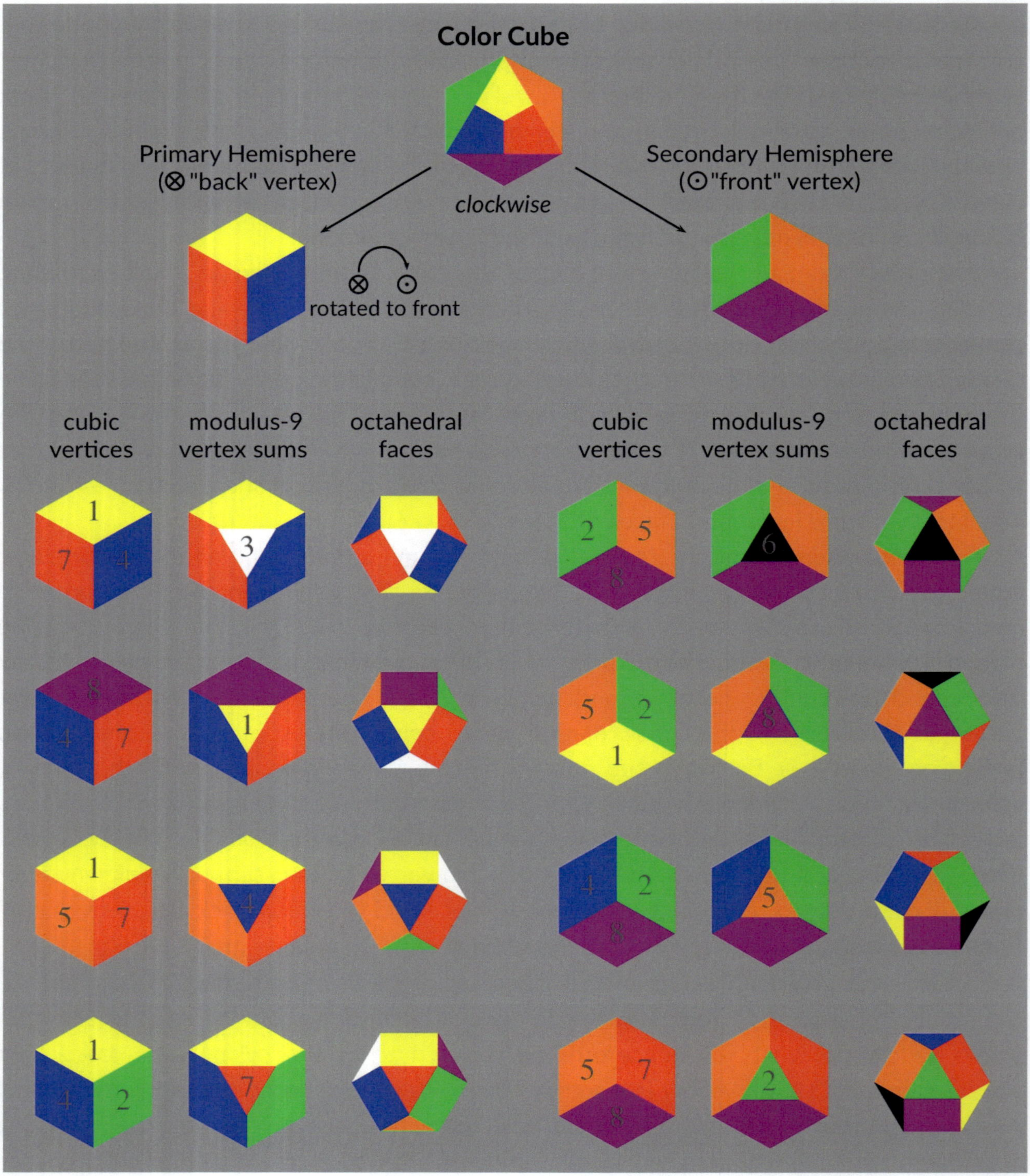

Figure 11. Cuboctahedral color-number polarity symmetry derived independently in two ways: either from a complementary cubic interpretation of published color symmetry or from the complementary cuboctahedral geometry of the 3-6 dipole highlighted through modulus-9 analysis of triangular number (Figure 10).

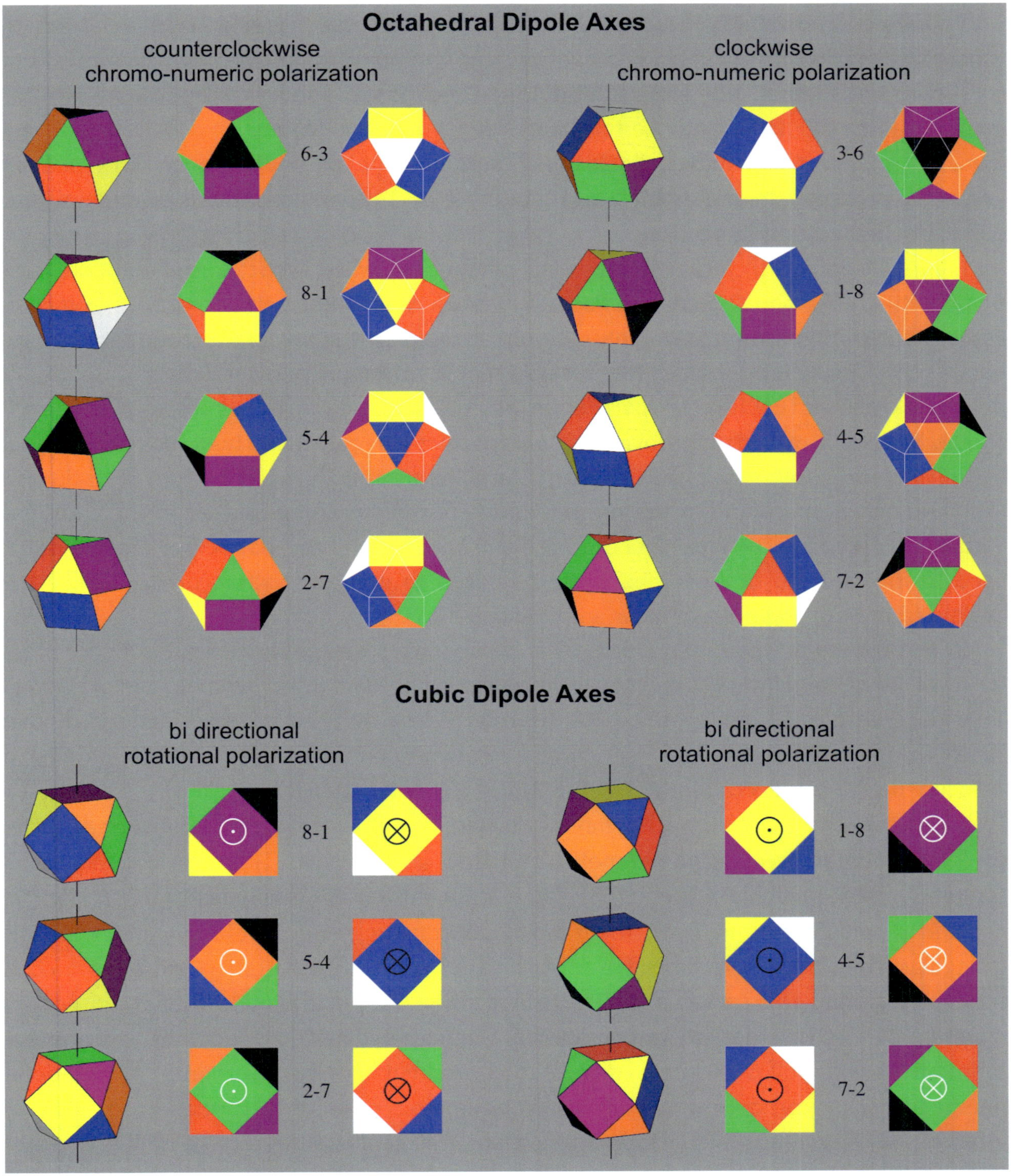

Figure 12. Complementary-cubic and complementary-octahedral dipole perspectives: counterclockwise cubic chromo-numeric polarization is shown on the left and clockwise on the right. Circle-dot means out of the plane of the page and circle-x means into the page as in previous figures.

Polarity-3 and polarity-6 octahedral faces suggest that adjoining face polarities of other cube vertices can be summed and reduced to find all octahedral face polarities; this is shown in Figure 11. Notice that since opposing cuboctahedral faces are all complements, all face and vertex axes of symmetry (cubic and octahedral) are complementary color-number dipoles that sum to nine or neutral grey. This polarity geometry produces "north" (white) and "south" (black) dipoles separated by neutral grey cuboctahedral equatorial vertices.

Figure 11 depicts a clockwise rotational perspective with respect to the numeric order of polarity (1-4-7 and 2-5-8), yet Pascal's Triangle has chiral symmetry. So a complete assessment of this form of symmetry should encompass the analogous counterclockwise perspective. Figure 12 provides the possible dipole rotational perspectives accounting for both clockwise and counterclockwise cubic polarization. A cuboctahedral elevation-view is shown to the left of each top-down axial perspective in Figure 12 to help the reader orient to a given axial-rotation perspective. For octahedral perspectives the bottom or back view is shown below an outline of the top or front faces versus being rotated to the front like Figure 11. Similarly, cubic perspectives are shown from the top down, but a circle-dot is used to indicate top or front and circle-x is used to indicate bottom or back since the outline of the top and bottom faces are the same.

Both octahedral and cubic dipole axial perspectives are provided in this figure since both will be needed in the coming analysis of the super-symmetric perspectives of Pascal's Triangle. These axial perspectives will be referred to by their chromo-numeric dipoles latter in this work. For instance complementary-cubic rotational super-symmetry is associated with rotation about either the 3-6 or 6-3 octahedral dipole axis.

Cuboctahedral color-number symmetry synergistically merges three concurrent forms of complementary symmetry: geometric, numeric, and chromic. Therefore I refer to this symmetry as *geo-chromo-numeric*. The same arrangement of complementary *geo-chromo-numeric* dipoles exists in both the monad coordinating 3-6 dipole symmetry of the triangular translational perspective and in the cubic interpretation of Itten's color symmetry model. This serves as a strong first affirmation of the validity of the proposed modulus-9 complementary-cuboctahedral super-symmetric geometry. Further affirmations will follow when I discuss the modulus-9 super-symmetry of rotational perspectives, in particular the continuous geometric proportion formed from rotation intervals of either spatial or temporal constraints, and continuous golden proportion formed from rotation intervals of either spatial/temporal or temporal/spatial constraints.

CHAPTER CITATIONS

[41] Fairbanks, Arthur 1898 *The First Philosophers of Greece* Charles Scribners (New York) p 134
[42] Iamblichus trans. Waterfield, R. 1988 *The Theology of Arithmetic* Phanes Press (USA) p 105
[43] Guthrie, K.S. 1988 *The Pythagorean Sourcebook and Library* Phanes Press (USA) p 28
[44] Iamblichus trans. Waterfield, R. 1988 *The Theology of Arithmetic* Phanes Press (USA) p 105
[45] Ore, Oystein 1948 *Number Theory and Its History* McGraw-Hill Book Co. (USA) p 19
[46] Ore, Oystein 1948 *Number Theory and Its History* McGraw-Hill Book Co. (USA) p 225
[47] Ore, Oystein 1948 *Number Theory and Its History* McGraw-Hill Book Co. (USA) p 210
[48] Ore, Oystein 1948 *Number Theory and Its History* McGraw-Hill Book Co. (USA) p 215
[49] Porphyry trans. Guthrie, K.S. 1988 *The Life of Pythagoras - The Pythagorean Sourcebook and Library* Phanes Press (USA) p 133
[50] Itten, Johannes 1961 *The Elements of Color* John Wiley & Sons (New York) p 31
[51] Itten, Johannes 1961 *The Elements of Color* John Wiley & Sons (New York) p 20-21

THE EXPANDING MOUNTAIN OF JEWELS

The Way gave birth to unity,
Unity gave birth to duality,
Duality gave birth to trinity,
Trinity gave birth to the myriad creatures.

The myriad creatures bear yin on their backs and
embrace yang in their bosoms.
They neutralize these vapors and thereby achieve
harmony.

Lao Tzu, Tao Te Ching [52]

The study of triangular number detailed in Chapter 5 confirms that there is a link between the dimensional symmetry of the continuous proportions of Pascal's Triangle and modulus-9. This affiliation between continuous proportions and modulus-9 symmetry stems from the geometric and numeric relationship between void (0) and point (1), which leaves an imprint on the numeric symmetry of dimensional perspectives that arise from the merger of these opposing symmetry elements. The resulting modulus-9 symmetry present in the continuous proportions of these dimensional perspectives provides a new means of interpreting dimensional geometry.

Modulus-9 highlights the complementary numeric harmony achieved when point and void merge to form dimensions. This harmony is achieved by neutralizing the opposing characteristics of point and void. Point and void are opposites in the sense that the point is finite (limited) and the void is infinite (unlimited). The central origin, the point, is inherently odd. Whereas the void that surrounds the origin is inherently even. The unlimited-even character of void orthogonally couples with the limited-odd character of the point to produce the mirroring continuous whole number proportions of Pascal's Triangle.

6.1 THE ANALOG CONTINUUM

Figure 13 is provided to help the reader visualize the reduction of Pascal's Triangle from modulus-10 continuous proportions to modulus-9 color-accentuated dimensional number polarities. An expanded view will reveal that this dimensional polarity field is composed of self-similar analogs of increasing complexity. The first nine-by-nine dimensional states comprise the first order analog of the dimensional polarity field since the linear dimension acquires its numeric invariance upon the differentiation of the number nine.

The first three by three dimensional states could be defined as the first order, but since they don't complete an analog cycle or *recurrence* culminating in the generation of a neutral polarity states, symmetry suggests they be considered a primordial order, i.e. one that gives rise to the flowering of the full range of first order states. The rational for this distinction should become clear later on when I correlate this polarity field with electron orbital theory (quantum numbers) and the periodic table of atomic elements.

1	1	1	1	1	1	1	1	1
1	2	3	4	5	6	7	8	9
1	3	6	10	15	21	28	36	45
1	4	10	20	35	56	84	120	165
1	5	15	35	70	126	210	330	495
1	6	21	56	126	252	462	792	1287
1	7	28	84	210	462	924	1716	3003
1	8	36	120	330	792	1716	3432	6435
1	9	45	165	495	1287	3003	6435	12870

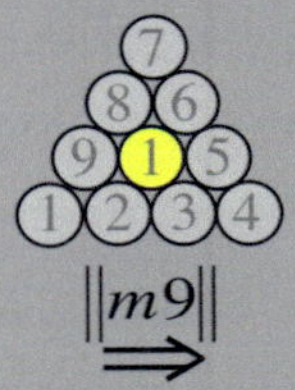

1	1	1	1	1	1	1	1	1
1	2	3	4	5	6	7	8	9
1	3	6	1	6	3	1	9	9
1	4	1	2	8	2	3	3	3
1	5	6	8	7	9	3	6	9
1	6	3	2	9	9	3	9	9
1	7	1	3	3	3	6	6	6
1	8	9	3	6	9	6	3	9
1	9	9	3	9	9	6	9	9

Figure 13. Modulus-9 transforms Pascal's Triangle into the dimensional number polarity field, which is then augmented using the complementary color-number affiliations defined in Chapter 5.

Examining a larger section of the field with Figure 14, where the primordial order (3x3), first order (9x9), and second order (27x27) field analogs are shown, gives a greater sense of the fractal-like self-similar symmetry under investigation. Neutral states mark the completion of analog polarity patterns that then repeat for each translational perspective. Notice however that the analog periodicity of translational perspectives is not fixed; rather it increases 3-fold with each succeeding analog order. For example, the translational analog period of the first three primordial dimensions is nine while the period of the first order dimensions is actually twenty-seven. A distinction must be made between perspective-specific analog patterns and the overall analog pattern of the field. References to primordial, first order, and so on pertain to the overall analog pattern of the field versus those of specific translational perspectives.

The analog periodicity of a given translational perspective can be seen by simply finding where the polarity pattern repeats. The neutral polarities are effectively forecast by the appearance of the three-six tertiary phase patterns in the preceding analog. This neutralization process begins with the primordial (3x3) analog tertiary phase pattern, which is neutralized within the first-order (9x9) analog. The primordial tertiary phase pattern is expanded or magnified to create the first-order tertiary phase pattern, which is subsequently neutralized within the second-order (27x27) analog field. This lag in the production of neutral polarities facilitates the symmetrical analog expansion of the field. Notice that the first-order field overlaps or contains the primordial, and likewise the second-order contains that of the first.

The number of tertiary phase polarities incorporated within dimensional perspectives increases with their degree of differentiation from the singular perspective. Also, beginning with the continuous arithmetic proportion, each translational perspective contains one neutral polarity per constraint at the end of each translational analog. So modulus-9 reveals the dimensional constraints discussed in Chapter 3. The compound dimensional constraints of each continuous translational proportion correspond to the number of repeating neutral polarities within each analog pattern. One invariant polarity is integrated into the analog translational polarity pattern for every constraint of void-to-point the perspective is subject to.

Notice, however, that the singular translational perspectives have no neutral species, and recall that both these perspectives are constrained solely by inversion. This means that the spatial translational perspective of the point is introverted and is composed of the continuous equal proportion formed by the continuum of possible temporal perspectives of the point. Likewise the temporal translational perspective of the point is extroverted and is composed of the continuum of spatial perspectives of the point. Further recall that the neutral void that surrounds the point also has continuous equal proportions. Point and void then, both having continuous equal proportions, are always self-same. Therefore both void and point are inherently un-merged or un-blended. The singular perspective has no neutral species and the perspective of void is entirely neutral. Only perspectives constrained by sums of void and point (merged perspectives) have both non-neutral and neutral species. And, as the point is repeatedly summed with void, there is an incremental progression toward increasing neutralization of the numeric species that form those merged perspectives.

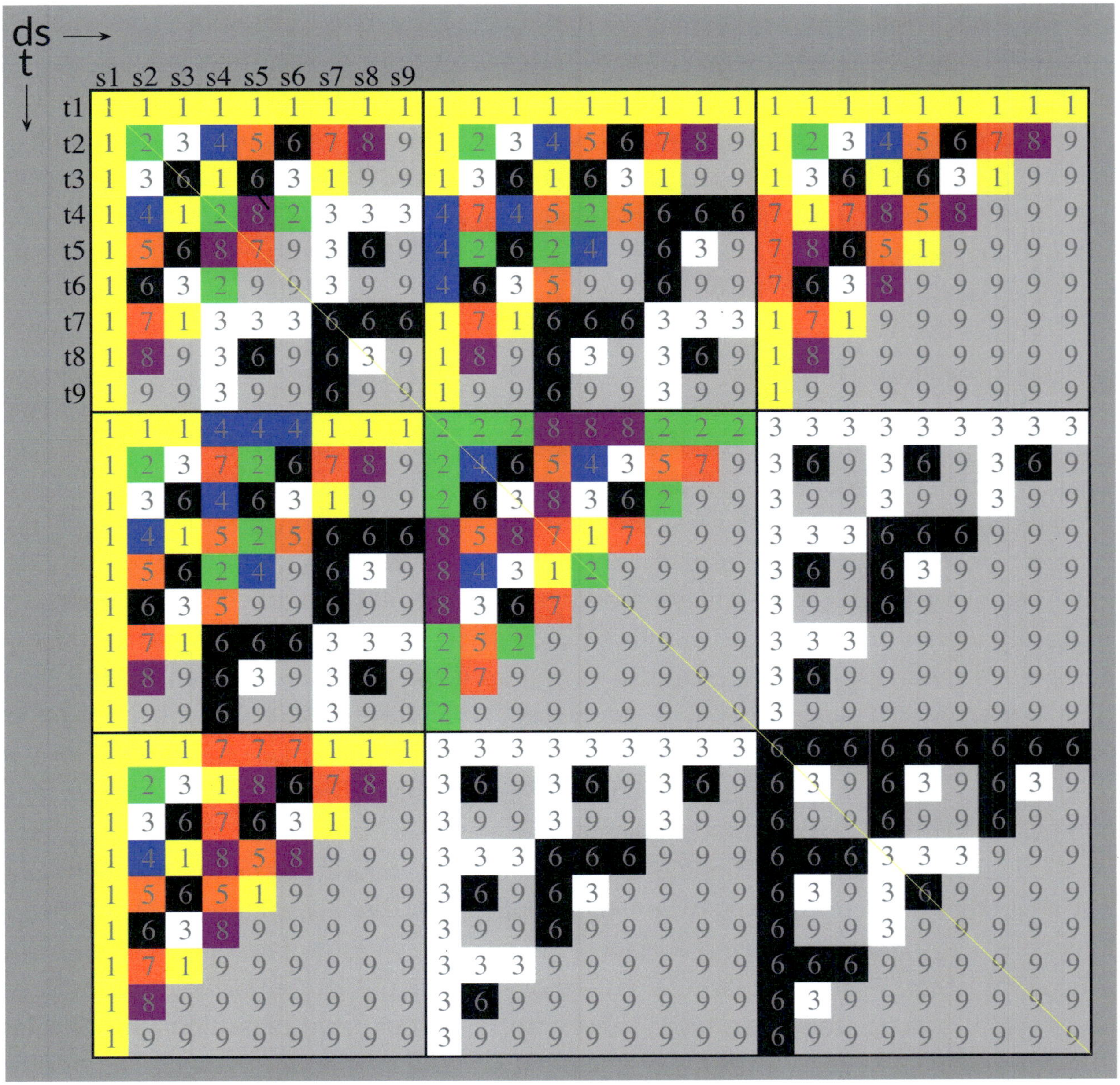

Figure 14. The Dimensional Polarity Field (27x27): The correlation between the number of dimensional constraints and modulus-9 invariance is most readily seen in the lower left and upper right quadrants where first-order translational analogs end with an incrementally increasing number of nines (1 to 8).

When point and void first merge modulus nine polarities are constrained to arithmetic order; therefore primary, secondary, and tertiary phases are splayed out such that every third number is primary, every third number is secondary, and every third number is tertiary. This arithmetic polarity pattern defines the first translational field analog. When merged, the orthogonal symmetry elements of void and point produce invariance in the continuous proportions that originate from them to the degree that the two symmetries are blended: Linear perspectives are differentiated by one constraint from singular perspectives and have one neutral polarity to mark the ends of their translational analogs (first blending). Triangular perspectives are differentiated by two constraints from singular perspectives and have two neutral polarities to terminate their translational analogs (second blending). Tetrahedral perspectives are thrice constrained and have three neutral polarities (third blending), and so on. From this analysis it is evident that modulus-9 reveals the underlying dimensional constraints active on continuous proportions that emanate from the combined perspectives of void and point.

Consider the effects resulting from the melding of these opposite symmetry elements: The field originates from the isotropic point and then, through first fusion with the surrounding void, the perspective of the point adopts the tangential perspective of void creating a form in the likeness of the tangent, the anisotropic dyad. But, recall from Chapter 3 that the geometric centers of all odd linear numbers coincide with the central geometric origin, the point. Whereas the geometric centers of even linear numbers do not contain a central point and instead contain a central void. So in this sense both the unlimited-even character of the void and the limited-odd character of the point are contained within the continuous arithmetic proportion of the line. Further, the continuous arithmetic proportion is unlimited in the sense that it is a continuous line and at the same time limited in the sense that it is composed of discrete points. This first fusing of opposites is the first incremental merging of the symmetric extremes of point and void. In effect the merging of opposites appears to *neutralize* their differences and in this way a *harmony* or balance is established between them.

As discussed in Chapters 3 and 5, the continuous triangular proportion highlights one of the geometric connections between continuous proportions and modulus-9 symmetry. Every third triangular plane (1-like planes) contains a central point coincident with the origin. The other 3-like and 6-like planes are centered about interstitial voids. And, these same planes produce the two neutral species at the end of each translational field analog reflective of the two constraints of void on point that the continuous triangular proportion is subject to.

Triangular number marks the end of the primordial order field. Beyond it modulus-9 symmetry highlights symmetric expansion of the primordial field symmetry. After the continuous equal proportion and the continuous triangular proportion, centered points again consistently coincide with polarity-1 at the end of the first order field (dimension 9) where every ninth geo-numeric species contains a central point (1st, 10th, 19th, …) and the other eight produce neutral species. The modulus-9 geometric pattern of the first 3 dimensions expands with base-3 geometric proportions every succeeding analog.

6.2 TIME-LIKE SPACE AND SPACE-LIKE TIME

As discussed previously, the next three constraints beyond tetrahedral dimensional perspectives fold back in opposition to preceding constraints (anti-parallel). For spatial dimensions this condition marks a return to introverted perspectives similar to the singular spatial perspective and the converse for temporal dimensions. These anti-parallel perspectives lead to the consideration of time-like spatial and space-like temporal dimensions – dimensions where temporally oriented continuous spatial translational proportions are time-like and dimensions where spatially oriented continuous temporal translational proportions are space-like.

Applying modulus-9 shows that the rotating constraint pattern of the linear, triangular, and tetrahedral spatial and temporal perspectives can be extrapolated to other dimensional perspectives since the cumulative invariance (neutral species) integrated within translational perspectives continues to increase by one for every constraint of void-to-point to which the succeeding perspectives are constrained to. This suggests that there is a continuum of orthogonal translational perspectives ordered by constraint orientation. This pattern is what allowed me to assign the initial constraint orientations of the singular spatial and temporal perspectives given in Chapter 3 and to infer that they are introverted and extroverted respectively, since only the linear, triangular, and tetrahedral constraint orientations are readily observable through the geometric figures detailed in that chapter.

Per Figure 15, the orthogonal juxtaposition of dimensional constraints results in six possible constraint orientations. Spatial and temporal constraints both rotate through all six possible orientations and are always anti-parallel to one another. Spatial constraints oriented such that they induce spherical close-pack space-filling geometric expansion of spatial forms are defined as *space-like*, as are those of the linear, triangular, and tetrahedral spatial continuous proportions. Temporal constraints that are *time-like* are anti-parallel to those that are defined as *space-like* and cause the negative reciprocal spherical close-pack time-filling geometric expansion of temporal forms, as do those of the linear, triangular, and tetrahedral temporal continuous proportions. Spatial constraints that are *time-like* are parallel to *time-like* temporal constraints, and temporal constraints that are *space-like* are parallel to *space-like* spatial constraints. By definition then, *Space-like* spatial perspectives are extroverted, and *time-like* spatial perspectives are introverted, and conversely *time-like* temporal perspectives are introverted and *space-like* temporal perspectives are extroverted. In other words an introverted spatial perspective is one that folds back upon a prior extroverted perspective and an extroverted temporal perspective is one that folds back on a prior introverted perspective.

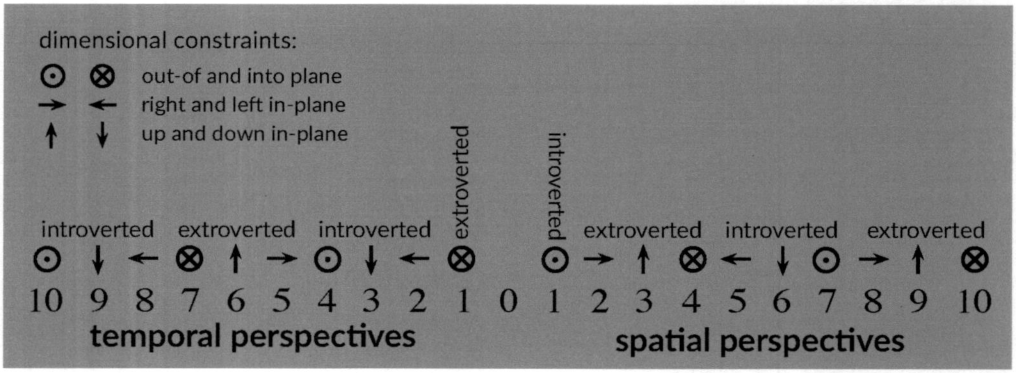

Figure 15. Extroverted (*space-like*) and introverted (*time-like*) dimensional constraints for spatial and temporal translational perspectives are shown relative to arithmetic proportion. Constraints are as defined in Chapter 3. The linear temporal perspective contains the second terms of all spatial translational perspectives and vice versa. These are the first terms to differentiate a given perspective from the singular. So I use the arithmetic proportion to show the relative constraint orientation of succeeding dimensions.

As stated, the six-fold rotating constraint pattern affirms that the singular translational spatial perspective is introverted and the singular translational temporal perspective is extroverted. This interpretation is consistent with the fact that the singular spatial translational perspective is the continuous equal proportion composed of all temporally constrained translational perspectives of the origin and that singular temporal translational perspective is the continuous equal proportion composed of all spatially constrained translational perspectives of the origin. Put more succinctly, the spatial translational perspective of the origin observes all its temporal selves and the temporal translational perspective of the origin observes all its spatial selves.

Notice that there is a unique correspondence between the six-fold rotation of dimensional constraints and the complementary cuboctahedral 3-6 dipole symmetry derived from the opposing 3-like and 6-like triangular planes of the continuous triangular proportion. The six-fold rotation between dimensional constraints causes the constraints of the spatial dimensions 3 and 6 to be anti-parallel just like spatial and temporal dimensions – the spatial dimension 3 is *space-like* and the spatial dimension 6 is its *time-like* complement. The orthogonal differentiation of translational perspectives naturally co-constrains (parallel constraints) 3-6 spatial-temporal and 3-6 temporal-spatial dipoles, and oppositely constrains (anti-parallel constraints) 3-6 spatial-spatial and 3-6 temporal-temporal dipoles. Further, co-constrained 3-6 dipoles pair *space-like* space with *space-like* time and vice versa. As such, they represent co-constrained complementary dipole perspectives that do not symmetrically neutralize one another since their triangular orientations are the same. This is like two octahedral faces with the same triangular orientation versus two that are in opposition such as the polarity-3 and polarity-6 faces of the same cuboctahedron. The unique position of 3 and 6 relative to constraint rotation gives 3-6 dipoles unique properties relative to other complements that will become apparent as the numeric symmetry analysis progresses. For instance, during the evaluation of triangular number I noted that 3-like triangular planes could have polarity-3 or polarity-6 and the same is true for 6-like triangular planes. Furthermore spatially oriented 3-like planes have the same spherical close-pack orientation as temporally oriented 6-like triangular planes and vise versa. I'll show that other complements are similarly paired with parallel and anti-parallel constraints, but only with regard to the geometry of super-symmetric rotational perspectives.

6.3 THE NUMBER OF THE ALL

> For as the Pythagoreans say, the All and all things are defined by threes; for end and middle and beginning constitute the number of the All, and also the number of the Triad.
>
> Aristotle [53]

Aristotle relays that the Pythagoreans taught that *all things are defined by threes*, since *the All* consist of a beginning, a middle, and an end. This defining characteristic of *the All* is a metaphorical reference to the properties of all continuous proportions, since they are all described by past, present, and future terms. The rotational pattern of dimensional constraints also consists of three constraints that are extroverted (space-like) and three introverted (time-like). Three polarity phases are embedded in modulus-9 and each consists of three polarities. Also, as previously discussed, the geometric correlation to modulus-9 is in the three spherically close-packed triangular planes that form the cuboctahedral coordination of the monad. It is unlikely that these 3-fold features are coincidental; in fact they all appear to be effects that arise from the ordering of number itself through its governing continuous proportions and the orthogonal juxtaposition of its two core symmetry elements – the infinite void (0) and finite point (1).

> Midway between the Being that is non-partitioned and always self-same [monad], and in turn the Being that is partitioned and comes to be in the realm of bodies [dyad], he blended out of both a third form of Being [triad] ...And once he had mixed them with Being and made one out of three, he again distributed this whole into so many portions as was fitting...
>
> Plato's Timaeus [54]

Plato suggests that after the formation of the triad from the fusion of monad and the dyad, a one was fashioned in its likeness. This *whole* was *again distributed* into the geometric proportions of the Platonic Lambda, which consists of powers of two and three emanating from the number one. The Platonic Lambda is derived from the Tetraktys and is associated with harmonic proportions since *the most perfect consonances are found in these numbers.*[55]

In Figure 16 I show the Platonic Lambda juxtaposed to the observable proportions of the dimensional polarity field. I do this in order to suggest that the numeric symmetry Plato was observing is not unlike that of the dimensional number field. Yet his Lambda misses the mark slightly.

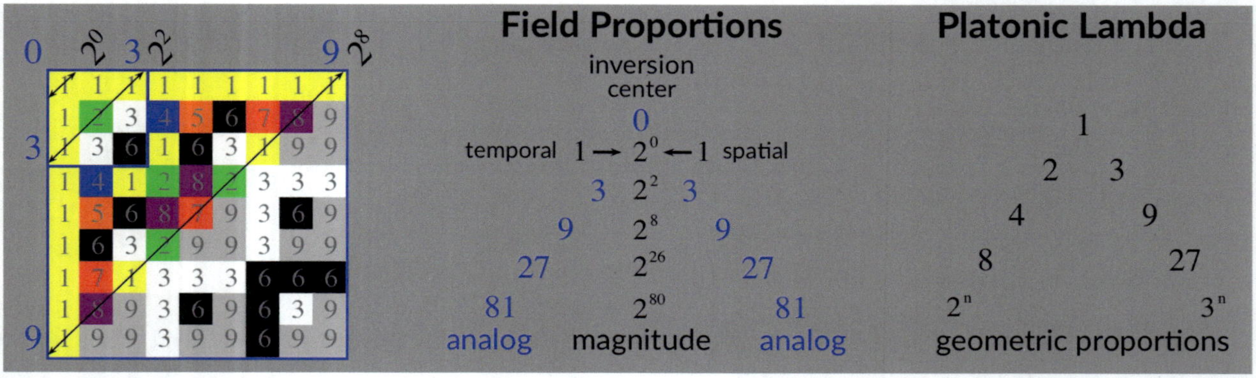

Figure 16. The complementary dimensionsal number polarity field versus Platonic Lambda proportions. Blue indicates field analogs and black indicates summed magnitudes of diagonals.

Plato seems to suggest that the melding of what is *always self-same* (i.e. the continuous equal proportions of the monad) with the other (the arithmetic proportions of the dyad) to form the triad causes both the 2 and 3 to become fused with *being* in the form of the monad which holds their natures within it seminally. Then, as the monad spawns all forms of geometric proportion, since, by convention any magnitude raised to the zero power is equal to one, both re-emerge to form parallel geometric proportions that compose the lambda.

Pascal's Triangle suggests that the convergence of base-2 and base-3 continuous geometric proportions at the number one is an artifact of the overlay of modulus-9 analog symmetry upon modulus-10 differentiated magnitudes. The sums of the diagonals (integrals) of Pascal's Triangle are in base-2 continuous geometric proportion. The base-2 proportion arises from modulus-10 dimensional magnitudes and the base-3 proportion arises from modulus-9 field analogs. This implies that the source of base-2 geometric proportion is the unlimited-even symmetry of the void that causes the limited-odd point to internally reflect with spatial-temporal mirror symmetry, whereas base-3 continuous geometric proportion appears more directly related to the numeric symmetry that arises from the constraint of void on point. So although Plato's observations are close to the truth, he attributes cause to an anthropomorphized *he* versus the reason (*logos*) or logic of dimensional number symmetry – *With Reason, not with hands, did the World-maker make the universal World.*

A study of some details resulting from the overlay of modulus-9 and modulus-10 follows. These details serve to further expound the interrelationship between these two moduli in the context of Pascal's Triangle and to highlight the symmetrical expansion of the primordial order analog. It should be noted however that this discussion is somewhat tangential to the main thrust of the book.

The number of neutral species terminating each equivalent spatial and temporal translational analog can be determined from the sum of the diagonal that connects them. When the diagonal sum is expressed in base-2 exponential form, the magnitude of the exponent equals the number of neutral species. For example, the triangular dimension has two spatial and two temporal neutral species that terminate each translational analog, which corresponds with the diagonal sum represented by two raised to the second power. This correspondence suggests that the base of two arises from the even mirroring of spatial and temporal perspectives and that its power arises from the number of dimensional constraints, beyond the singular, imposed on said perspectives.

Looking now to the coincidence of the powers of two with the base-3 continuous geometric expansion of field analogs one sees that starting with the primordial order three, every analog ends with a diagonal magnitude that reduces to four. The reason for this is, beyond the primordial order, these diagonals are composed entirely of the complements 1+8 and 2+7 and that they always contain two 1s and one 2 that do not neutralize when the entire diagonal is summed and reduced, just like the primordial order diagonal. This is due to the mirror symmetry of the diagonal that clusters complements into odd numbered groups (3, 9, 27, ...) along the diagonal and mirrors two 1+8 sub-fields around a central 2+7 field. This same symmetry is the source of the tertiary polarity pattern of threes that emanates from the confluence of the 1+8 and 2+7 subfields.

Taking a broader view, with Figure 17, note the beauty of the field reminiscent of a self-similar fractal-like pattern first observed by Sierpinski[*] with modulus-2. The various overlapping analogs are most evident in the central green sector of each analog beginning with the dyad. Observe the various seminal features present in the primordial 3x3 analog that produce the ever-increasing complementary complexity of subsequent analogs.

---

[*]Sierpinski's triangle is a modulus-2 version of Pascal's triangle that shows a similar pattern stemming from the distribution of odd and even numbers in the field.

Notice that the first three spatial and temporal perspectives culminate with the triangular numbers 1, 3, and 6, which are indicative of spherical close packing and modulus-9 symmetry. Also notice that the primordial order field analog contains the minimum of three terms required to define a proportion, and contains all six orthogonal constraint directions.

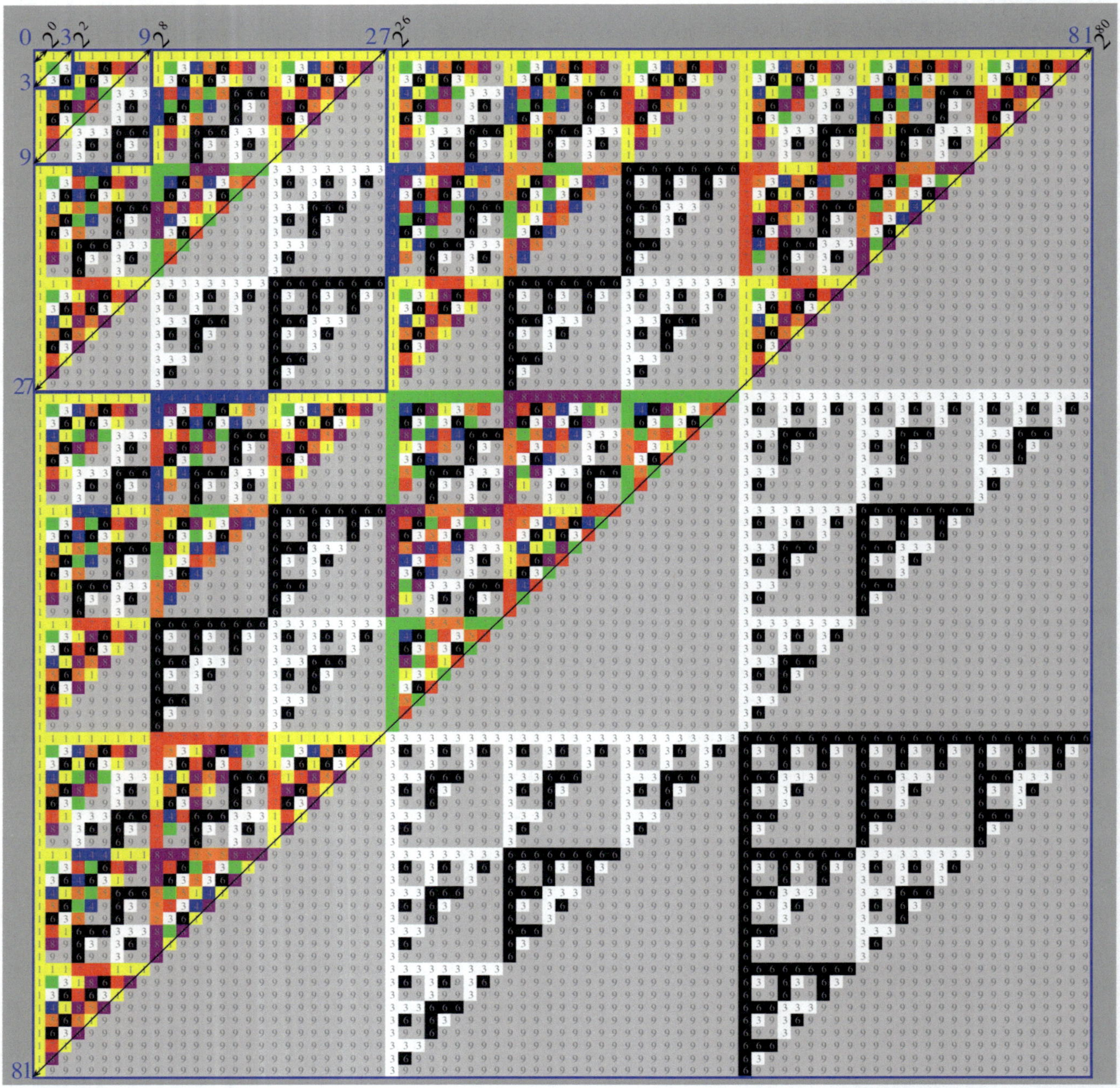

Figure 17. An expanded view of the dimensional polarity field showing the following overlapping analog periods: 3x3, 9x9, 27x27, and 81x81 (each is outlined in blue).

CHAPTER CITATIONS

---

[52] Lao Tzu trans. Mair, V.H. 1990 *Tao Te Ching (Ma-Wang-Tui Manuscripts)* Bantam Books (New York) p 9
[53] Fairbanks, Arthur 1898 *The First Philosophers of Greece* Charles Scribners (New York) p 134
[54] Plato trans. Kalkavage, Peter 2001 *Timaeus* Focus Publishing (USA) p 64
[55] Theon of Smyrna trans. Lawlor, R. and D. 1979 *Mathematics Useful for Understanding Plato* Wizards Bookshelf (San Diego) p 63

Complementary Numeric Transformations

All of the continuous proportions of the rows and columns of Pascal's Triangle are governed by translational symmetry transformations. Each of these forms of translational proportion is related to the other through the *continuous space-time proportion equality* (equation 4.2.1). The geometric forms governed by these proportions grow through translation along a given axis of constraint that is orthogonal to the preceding dimensional constraint, such that the species of the preceding proportion may be integrated, one upon the next, in the direction of the succeeding translational constraint. In contrast to translational proportions, the summed or integral proportions of Pascal's Triangle are governed by rotational symmetry transformations. For instance, I'll show that the continuous geometric proportions of the sums of sequential diagonals are in complementary-cubic proportion with respect to rotation about the 3-6 dipole axis of the complementary color cube. The geometric symmetry of these rotational proportions is associated with the rotation of constraint about a given axis between translational perspectives in contrast to the constrained translations along lines of sight that are indicative of translational proportions. It is informative to study the complementary modulus-9 numeric symmetry transformations that accompany these two categories of proportion by comparing and contrasting continuous arithmetic and geometric proportions.

7.1 Arithmetic Translation

The field of integers generated with continuous arithmetic proportion can be considered the media within which all continuous proportions propagate, since the linear dimension alone contains all integers. So in an analogous sense the arithmetic proportion is like the backbone of number that supports all forms of continuous proportion and their accompanying modulus-9 symmetry transformations. Hence, studying the basic linear translations of moduli along the backbone in the context of complementary numeric transformations provides a more comprehensive understanding of some of the basic characteristics of numeric complements.

In order to evaluate arithmetic translation or movement along a number line in terms of complementary symmetry transformations it is important to remember that, through scaling or multiplication, movement can range from increments of 0 to *n* since the constraint axis of arithmetic proportion is continuous. In order to observe these polarity transformations a range of scaled arithmetic translations in multiplication table form will be examined to understand translations not only by multiples of one, but two, three, four and so on.

To produce the scalar polarity transformation matrix, integers are arranged in continuous arithmetic proportion. Then these numbers are scaled systematically with the same arithmetic proportions and reduced to modulus-9 polarities. The result is an arithmetically ordered multiplication matrix that has then been reduced. Figure 18 shows how the scalar polarity transformation matrix is produced. Notice that the polarity patterns of all columns and rows repeat every nine integers. Therefore a 9x9 unit cell can be used to describe the complementary numeric symmetry transformations of the entire continuous analog scalar polarity field.

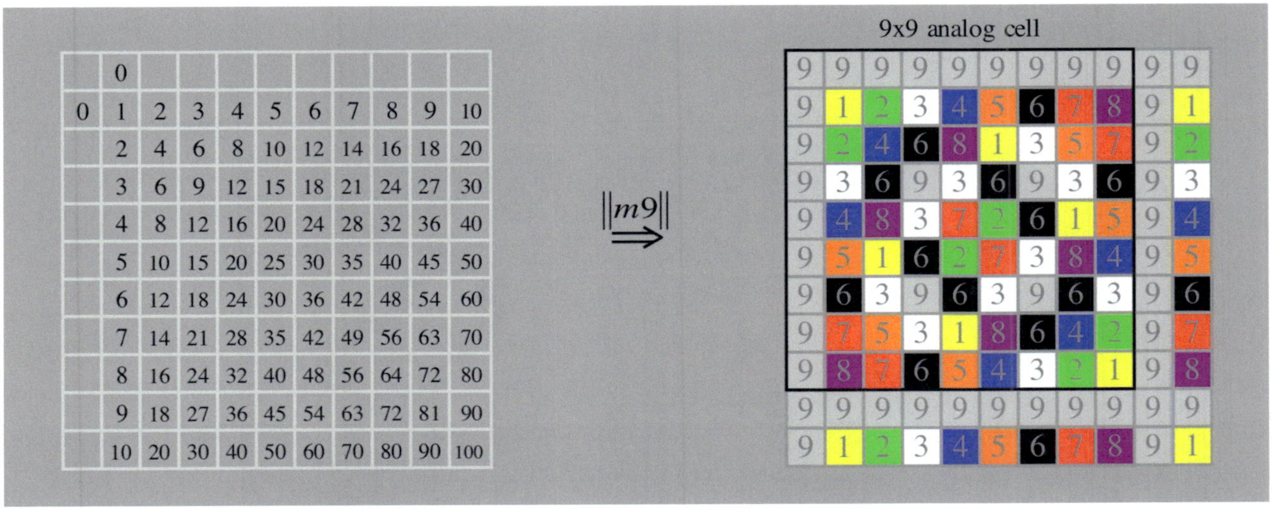

Figure 18. Scalar polarity field produced through modulus-9 reduction of a multiplication matrix. The 9x9 symmetry pattern (analog cell) repeats continuously throughout the scalar number field.

Arithmetic translations are achieved through polarity transformations that are opposite for modulus-9 complements. Polarities 1 through 4 cause precession of polarity and polarities 5 through 8 cause recession of polarity.* Generally speaking, the scalar polarity field shows that polarities symmetrically transform to their complements through a series of eight transformations, then they are neutralized at the ninth, and revert to their original polarity at the tenth translation, with the exceptions of the 3 and 6 which have a three times greater analog frequency, and of course the unchanging invariant monopole 9. All of these transformations occur along the same continuous number line. So with translations of 2 for example one is just observing even numbers such as those from 0 through 18 or 18 through 36.

Examining the unique character of each translational polarity inversion with Figure 19 shows that 1 translates through one objective group and one subjective group to invert to 8; 2 translates through two objective and two subjective groups to invert to 7; 3 translates through three of each to invert to 6 three times; 4 translates through four of each to invert to 5. Their complements undergo the same inversion in the opposite direction. Notice that integer multiples of 4 completely disperse objective and subjective polarities and reverse their arithmetic order relative to the initial ascending arithmetic polarity sequence. Inspection of the progression from the initial arithmetic polarity transformation sequence (2nd row or column) to scaled transformations by multiples of 4 (5th row or column) suggests a stepwise inversion between ascending (1, 2, 3, 4) and descending (4, 3, 2, 1) forms of polarity transformation, with inversion between the two occurring at the tertiary phase (4th row or column).

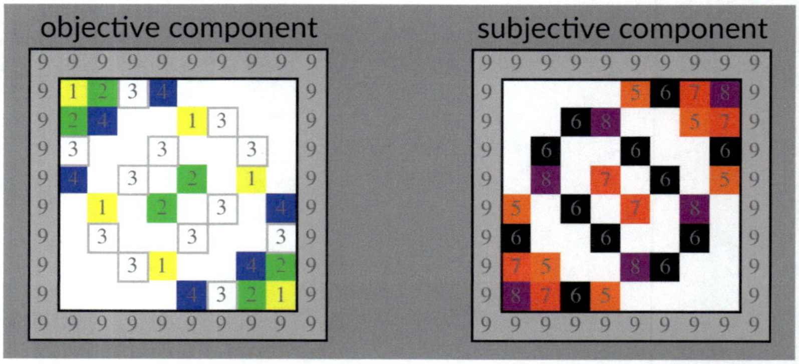

Figure 19. Complementary polarity transformations split into *objective* and *subjective* components: this analysis highlights the unique way each polarity transforms to its complement.

---

*Recall from the Chapter 5 number wheel example that *objective* polarities cause precession and *subjective* polarities cause recession in the modulus-9 residue.

Another way to analyze scalar symmetry is to separate polarities by phase (Figure 20). This analysis highlights the 3-fold interleaving of primary, secondary, and tertiary phases within the scalar polarity field:

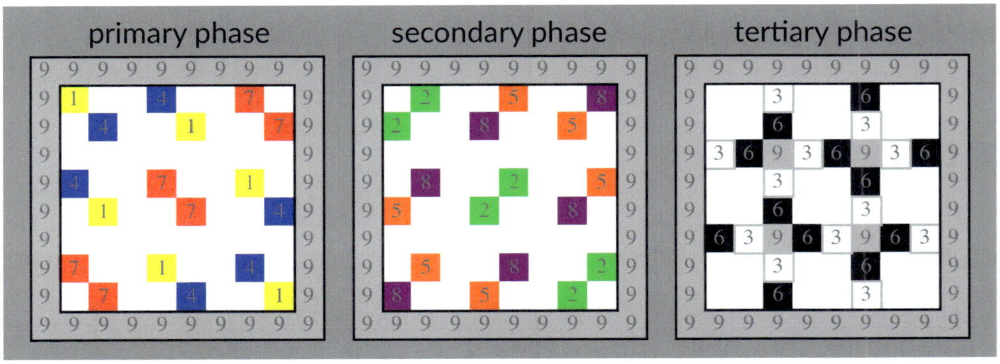

Figure 20. Primary, secondary, and tertiary phase distribution within the scalar polarity field.

The preceding analysis shows that scalar translation is constrained to the axis of the continuous arithmetic proportion regardless of the size of the steps taken along its constraint axis since these scaled arithmetic proportion variants all exhibit complementary transformations culminating in the production of an invariant polarity. Scalar steps larger than one simply bypass segments of the number line. All such scalar proportions, being constrained to the linear perspective, are therefore governed by its complementary numeric symmetry.

It's interesting to consider whether these complementary numeric transformations, unique to each complementary pair, are in some way indicative of their larger roles within the media of the dimensional construct since for any system to function in a predictable and orderly manner it seems natural for the participants in the system to have unique and indispensible roles. And since modulus-9 governs all numeric operations through a set of nine natural numbers, it seems reasonable to presume then that each must have such a unique role. And if these roles were not already established and coordinated about the singular primordial state, as complementary-cuboctahedral dipoles perhaps, it's difficult to comprehend how else the singular state would harbor the fundamental symmetries necessary to sustain natural order.

To more fully understand these fundamental symmetries we next turn to the complementary rotational transformations that accompany continuous geometric proportions. Recall that translational perspectives emanate outward from a central origin *along* axes of constraint. In contrast rotational perspectives revolve about the same central origin *between* axes of constraint. I refer to these rotational perspectives as super-symmetric because they arise from the inherent or predisposed spatial-temporal symmetry of the point, which I envision as the super-position of two chiral symmetries associated with cuboctahedral spherical close packing symmetry, as shown in Figure 12. To begin I'll evaluate the cubic aspect of this symmetry in the next section.

7.2 GEOMETRIC ROTATION

> Some...agreeing with Philolaus...say that 'geometric harmony' is the cube because it is harmonized in all three dimensions, being the product of a number thrice multiplied together.
>
> Nicomachus* [56]

In mathematics the terms "squared" and "cubed" are legacies from ancient number theory. The term geometric proportion was coined to describe what was seen as the *geometric harmony* of the cube. The side of a cube can be considered the base of an exponential power series; when the base is multiplied by itself once it forms a square (area), and multiplied by itself twice it forms a cube (volume), the square area being produced from two orthogonal sides and the cube volume from a third. If the proportion is continued beyond the third term, it moves beyond a simple cubic analogy (even though any three consecutive terms can be represented by a cube). It follows that sequential terms of a given geometric proportion are orthogonal to one another since the present power of the base is always orthogonal to both former and future powers the base. Notice that this condition is consistent with the continuous geometric proportion of the diagonals of Pascal's Triangle since sequential diagonals unfold by rotation through the same six-fold pattern of orthogonal dimensional constraints that also define the relative orientation of translational perspectives.

Figure 21 shows the complementary symmetry of the field of continuous geometric proportions. This figure was constructed by starting with an arithmetically ordered column of integers and then subjecting it to sequential exponential transformations in order to create geometrically proportioned rows that are then reduced to modulus-9 color accentuated complementary polarities. Extra rows and columns are provided to show that the symmetry of the entire ordered field of continuous exponential proportions may be described by a repeating pattern of six modulus-9 polarity transformations; this 6x9 field repeats continuously throughout the ordered medium of whole numbers.

Curiously, there is a regular periodicity of bands of polarity-1 that seem to be symmetric repetitions of the initial row associated both with the number 1 raised to any power (every nine rows) and anything raised to the zero power associated with the initial column (every six columns). Yet in the latter case of the initial column, bases of the tertiary

---

*Nicomachus of Gerasa was Platonist writer in the 2nd century CE.

phase (3, 6, 9, ...) become neutral beyond their initial base term and are not associated with polarity-1 other than the initial boundary condition that any number raised to the zero power is equal to 1. Hence the vertical bands of polarity-1 every six columns are broken up by superseding horizontal polarity-9 bands every three rows. This artifact of the polarity field appears to be connected to the inherent infinite continuous equal nature of the number 0 which is reflected by the initial neutral horizontal polarity band.

n	n^0	n^1	n^2	n^3	n^4	n^5	n^6	n^7
0	?	0	0	0	0	0	0	0
1	1	1	1	1	1	1	1	1
2	1	2	4	8	16	32	64	128
3	1	3	9	27	81	243	729	2187
4	1	4	16	64	256	1024	4096	16384
5	1	5	25	125	625	3125	15625	78125
6	1	6	36	216	1296	7776	46656	279936
7	1	7	49	343	2401	16807	117649	823543
8	1	8	64	512	4096	32768	262144	2097152
9	1	9	81	729	6561	59049	531441	4782969
10	1	10	100	1000	10000	100000	1000000	10000000
11	1	11	121	1331	14641	161051	1771561	19487171

$\|m9\|$ $\Rightarrow$

6x9 analog cell

9	9	9	9	9	9	9	9
1	1	1	1	1	1	1	1
1	2	4	8	7	5	1	2
1	3	9	9	9	9	9	9
1	4	7	1	4	7	1	4
1	5	7	8	4	2	1	5
1	6	9	9	9	9	9	9
1	7	4	1	7	4	1	7
1	8	1	8	1	8	1	8
1	9	9	9	9	9	9	9
1	1	1	1	1	1	1	1
1	2	4	8	7	5	1	2

Figure 21. Formation of the polarity field of continuous exponential proportions: The figure demonstrates that a 6x9 exponential polarity field repeats continuously throughout the arithmetically ordered whole number field. Tertiary polarities of the first and second analog are underlined for comparison since they differ between the first and all subsequent analogs.

Conventionally the zero power of zero is either treated as undefined or as equal to 1 so the condition is represented with a question mark in the upper-left of Figure 21. However, modulus-9 symmetry suggests that, beginning with zero, tertiary polarities raised to the zero power should be neutral (polarity 9) instead of 1, since in subsequent six-fold analog periods their polarity states are all neutral. Compare, for instance, the underlined terms of the first and second field analogs.

This paradox is the crux of a long-standing mathematics question: What does zero to the zero power equal? On the one hand we have continuous geometric proportion in which the first term is to the second as the second is to the third and so on, which affirms that in all cases other than zero the initiating term should be 1 – since the initial ratio that defines the base of a given continuous geometric proportions is the base over 1. Yet this logic cannot be applied to the geometric proportions of zero because it would equate the ratio 0 over 1 with 0 over 0 (as shown in Chapter 4, the ratios of any two consecutive terms of a given geometric proportion define its base). This equates 0 with 1, whereas prior analysis suggests that the number 1 represents the finite geometry of the point and the number 0 represents the infinite geometry of the void that surrounds it. It would seem then that the logic of attempting to subject the infinite void to exponential growth is inherently flawed – Only something that is finite can grow through geometric replication of itself (exponentially or otherwise).

The analysis of Pascal's Triangle in Chapter 6 demonstrated that there is only one form of continuous geometric proportion originating directly from the point: The continued doubling in magnitude of the sums of its diagonals. The continuous proportion of these succeeding diagonal sums is constrained by the rotation between orthogonal orientations of succeeding continuous translational proportions. Similar to translation, where scaling can occur in steps of various sizes, rotation can occur in larger increments and still be said to be rotations through the same field of continuous geometric proportion. For example: rotation between orthogonal constraints generates the series 1, 2, 4, 8, 16, 32, 64 and so on (see row n=2 of Figure 21). If the amount of each rotation is doubled, then every other number in the series is skipped (1, 4, 16, 64, and so on, see row n=4). If the amount of each rotation is tripled, then two out of three numbers in the series are skipped (1, 8, 64, and so on, see row n=8). So, in this example, one can say that the latter two series are subsets of the former and that therefore they all originate from the same central origin.

The remaining geometric proportions together with the former comprise the 6x9 polarity field, and those that remain share a common symmetry with base-2, in that squaring and cubing any given magnitude forms such proportions. Therefore they all share a common 6-fold polarity transformation symmetry.

However, the complementary symmetry transformations that accompany constraint rotations between the diagonals of Pascal's Triangle are cogent here, as their symmetry transformations correspond to the symmetry of the complementary color cube discussed in Chapter 5. Per Figure 22, the complementary polarity transformations of base-2 continuous geometric proportion correspond to 6-fold rotations about the 3-6 dipole axis (cube diagonal) of the complementary color cube. All color cube polarities are orthogonal to one another consistent with the dimensional constraints highlighted previously, though rotation between them occurs in 60° increments. These 60° rotations correspond with polarity transformations that alternate between primary and secondary phase hemispheres of the color cube. After six rotations about the 3-6 dipole axis (360°) the polarity pattern repeats itself.

This 6-fold cubic pattern can be examined analogously by studying the continuous geometric transformation of the 30°-60°-90° triangle about a central axis, as shown in

Figure 22. Numeric transformations are achieved with this triangle geometry by equating the length of the hypotenuse of a given triangle with the length of the short side of the next triangle in the geometric series, which is rotated 60° with respect to the former. Therefore, a rotation of 60c corresponds with a doubling of the length of the hypotenuse, since the length of the hypotenuse is twice that of short side and since the angle between them is 60°; this further corresponds with a continuous doubling in the area of concentric circles whose radii are equal to the geometrically proportioned hypotenuses.

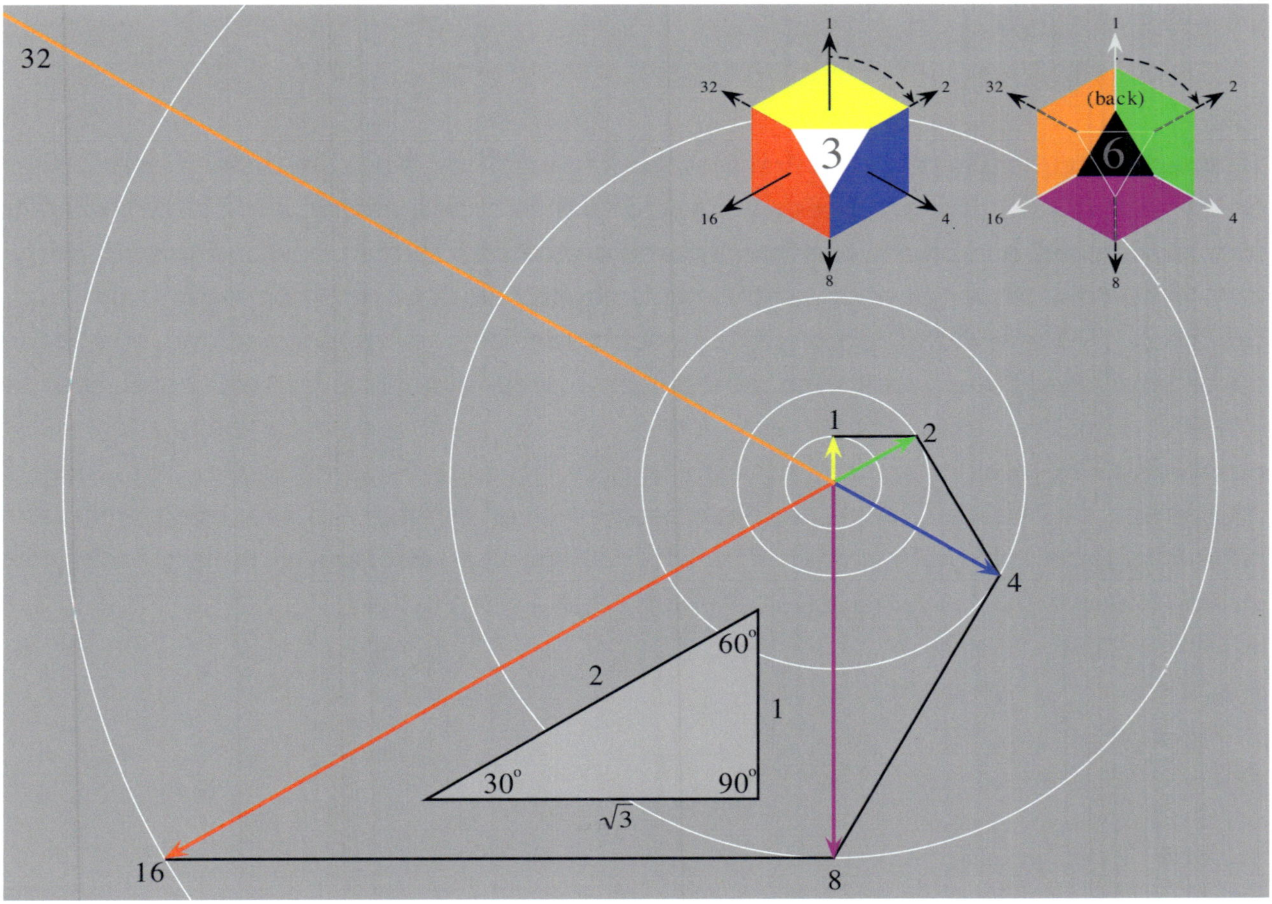

Figure 22. Base-2 continuous geometric proportions represented with 60° rotational transformations of 30°-60°-90° triangles about a central axis. The hypotenuse of a given triangle is the short side of the next larger triangle in the series. Equivalent color cube transformations are shown in the upper right to highlight alternation between primary and secondary phase hemispheres with 60° rotations about the 3-6 dipole axis.

In Philolaus' teaching continuous geometric proportions are formed when the length of the side of a given cube is *squared* and *cubed*. Yet here the geometric model is different, since base-2 is now associated with the ratio of the hypotenuse to the short side of the 30°-60°-90° triangle instead of the ratio of the face-area to side-length of a given cube.

Geometric growth in magnitude is thus associated with cube-face polarity instead of with the discontinuous cubic proportions associated with the ratios of length to area and area to volume. This leap can be a source of confusion, since it is obvious that in the case of a simple cube the area of all faces are equal, which begs the question: How is it that a symmetry can be cubic without all faces being equal?

> If as many numbers as we please beginning from an unit be in continued proportion, and the number after the unit be square, all the rest will also be square. And if the number after the unit be cube, all the rest will also be cube.
>
> Euclid's Elements Book Nine: Proposition Nine [57]

Imagine a cube where on the first face the lengths of the sides are defined as one, then on the second face two 1x1 faces are stacked so they appear as one, and on the third face four are stacked, and on the fourth eight, and on the fifth sixteen, and on the sixth thirty-two. In so doing super-positioned stacking is equated with integral summation, and cubic symmetry is preserved. The polarity of each cube face corresponds to the number of super-positioned faces (point symmetries) represented in modulus-9. And, the succeeding integral sums of these point symmetries are in continuous complementary-cubic proportion. The notion of super-positioning is crucial to understanding the finite integral proportions of Pascal's Triangle and their geometric correspondence with complementary-cubic and cuboctahedral symmetry transformations. It is also crucial understanding the virtual dimensional symmetry states of energy that govern matter.

CHAPTER CITATIONS

[56] Nicomachus of Gerasa trans. D'Ooge, M.L. 1926 *Intro. to Arithmetic* MacMillan (New York) p 277
[57] Euclid trans. Heath, L. 2002 *Euclid's Elements* Green Lion Press (USA) p 216

FIELDS OF CREATION

So far the path of reasoning has yielded a dimensional number theory, as yet incomplete, that suggests space-time states differentiate geometrically from a singular state in accordance with the dimensional constraint symmetry of Pascal's Triangle. I have deliberately left aside the fact that a point has no size until now. When a size-less point is internally reflected off of the spherical void that surrounds it, a continuum of virtual states is established, but no emptiness is enveloped.

As an analogy, visualize the case of two opposing mirrors with you as the point centered between them: peering into either looking glass you see a continuum of virtual replicas of yourself, this is like the linear dimension. Then take the analogy a step further and visualize another set of opposing mirrors positioned at right angle to the first set that reflects the reflections of the first set perpendicular to the first line of sight, this is like the triangular dimension. Take this one step further with a third set of opposing mirrors orthogonal to the prior two sets and you have an analogy for the perspective of the tetrahedral dimension where central point is cuboctahedrally coordinated by a continuum of virtual replicas of itself. The origin is thereby differentiated and integrated upon itself to form the geometric states of Pascal's Triangle without enveloping any of the surrounding void — all symmetry states are virtual.

This interpretation is consistent with the cuboctahedral rotational super-symmetry of the central point which suggests that, even though translational perspectives are differentiated by their constraints and by their associated continuous proportions, their differentiated states are all still super-positioned upon the same point. This is because they sum or integrate to form the complementary polarities that are coordinated about the origin. The monad, then, is present seemingly alone in the void. Yet with all its diverse states it appears to harbor the seminal capacity to sustain the diversity of material forms. Ultimately I'll show that Pascal's Triangle represents the continuum of virtual quantum energy states, a complementary spatial-temporal quantum number field.

This super-positioning of spatial-temporal translational symmetries produces the super-symmetric continuous proportions of Pascal's Triangle that I believe gives rise to the complementary dipole fields we observe from our matter-centric perspective as electrons and protons and as their anti-*particles* as well. Yet we'll be evaluating these fields first from a virtual energy symmetry perspective in order to understand how they appear as they do from the matter perspective.

Matter in its most fundamental atomic form, Hydrogen, is formed when a single electron is bound to a single proton. This bonding consumes a localized region of void that constitutes the bulk of the volume of said atom. In fact electrons and protons are often modeled as point-charges. In a sense, then, it is the formation of these localized bundles of energy (point-charge *particles*) and the integration of this empty space that distinguishes matter from energy. Since energy dimensions are composed of virtual states no void is enveloped through their formation. And the super-symmetric energy symmetry transformations that govern the evolution of these virtual states are what give rise to these point-charge *particles*.

8.1 Super-symmetric Transformations

Spatial and temporal constraints define finite intervals of super-positioned dimensional symmetry states that participate in complementary symmetry transformations, such as the summed proportions of the ordinary 1:1 diagonals of Pascal's Triangle. The finite integrals represented by these 1:1 diagonals have complementary-cubic rotational super-symmetry that can be either spatial or temporal depending on whether finite intervals are distinguished by spatial or temporal constraint rotation. These super-symmetric proportions have a single exponential base.

The shallow 2:1 diagonals of Pascal's Triangle have complementary cuboctahedral rotational super-symmetry that is either spatial/temporal or temporal/spatial. These super-symmetric continuous proportions have both a base and an anti-base owing to the spatial-temporal complexity of the interwoven dimensional constraints that define their transformation intervals.

Since Nature evidently prefers a spatial/temporal orientation, it makes sense that this orientation should be associated with the order in which dimensional constraints are imposed. It follows then that the constraint imposed on spatial differentiation should precede that of temporal differentiation for each translational perspective. This logic is used when the 1:1 and 2:1 (spatial:temporal) constraint intervals of Pascal's Triangle are defined in the coming sections.

8.1.1 COMPLEMENTARY-CUBIC ROTATIONAL TRANSFORMATIONS

The intervals of the super-symmetric spatial rotational perspective are defined by incremental temporal constraint imposition similar to the way spatial translational perspectives are formed with temporal increments. The notion of incremental constraint imposition is consistent with the continuously equal proportions of singular spatial and temporal translational perspectives being formed by successive rather than by concurrent sums of void and point. Figure 23 shows how incremental temporal constraint imposition is accounted for to define super-symmetric spatial rotational intervals, wherein temporal translational perspectives are shifted to the right (spatially) by one position for every additional dimensional constraint imposed upon them. As noted before, this integrated invariance results from the constraints imposed by succeeding sums of void and point. So integrated invariance can therefore be used as a means of accounting for incremental constraint imposition as shown in Figure 23.

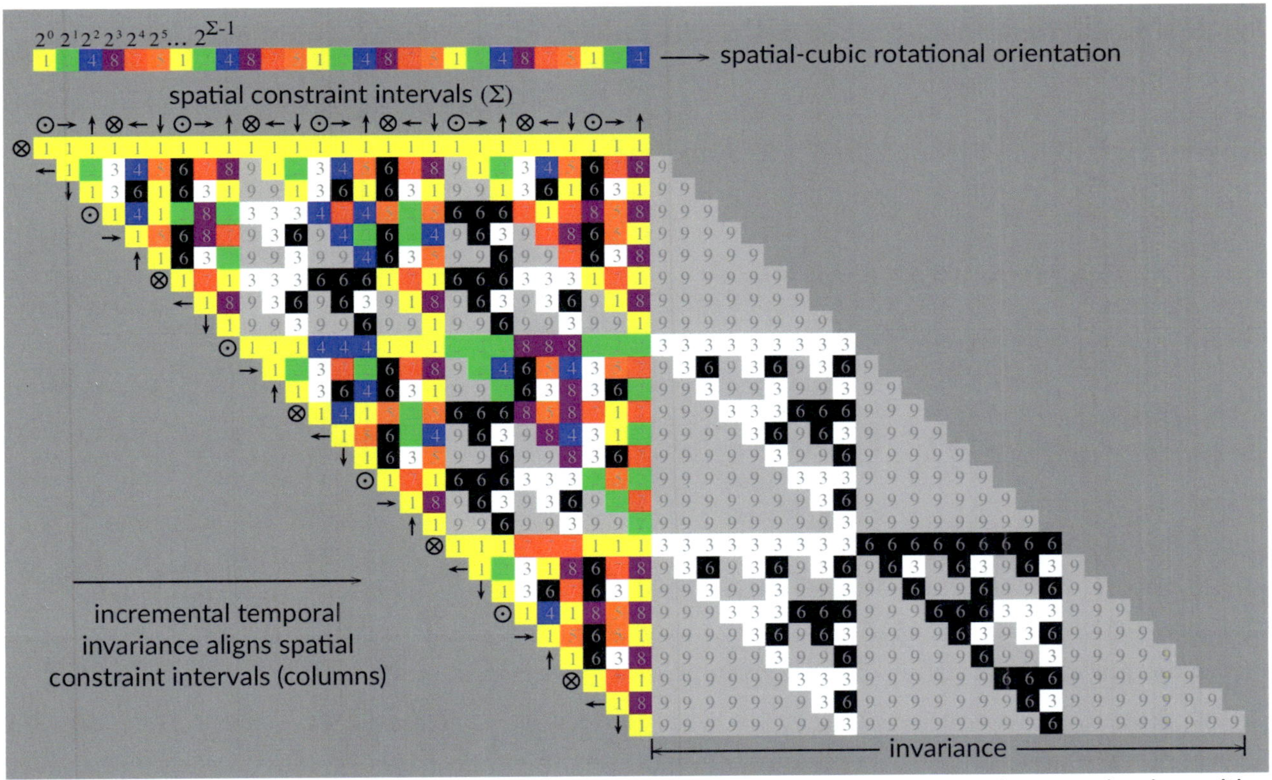

Figure 23. Super-symmetric spatial intervals: accounting for incremental temporal constraint imposition aligns column-intervals. This is shown to correspond with cumulative temporal translational invariance in the right hand portion of the figure. The sums of columns are in base-2 geometric proportion, which is accompanied by clockwise complementary-cubic symmetry transformations as shown in Figure 22.

Spatial intervals occupy columns that, when summed, are in continuous clockwise complementary-cubic proportion – this is the super-symmetric spatial rotational perspective. If the anti-perspective were adopted then temporal integrals would be in rows that are in counterclockwise continuous complementary-cubic proportion. This confirms that the continuous geometric proportions of integral spatial and temporal rotational perspectives arise from their respective constraints alone rather than from their combination. In other words these proportions reflect symmetry transformations between either spatial or temporal intervals, rather than between either combined spatial/temporal or combined temporal/spatial intervals.

Finite integral mathematics can then be leveraged to define the constraint intervals shown in Figure 23. Equivalent spatial intervals are defined in equation 8.1.1 below. These integrals correspond to the column intervals, which are created by shifting the dimensional polarity field graphically. Hence the reader is not obliged to decipher equation 8.1.1 since it is merely a mathematical representation of what is more easily comprehended graphically with Figure 23. Equation 8.1.1 is provided as an alternative means of comprehension for those so inclined since it puts the previously mentioned super-symmetric intervals in *integral* form thereby explaining why I refer to them as integrals.

Let $2^{\Sigma-1}$ signify the integral of spatial magnitudes (${}^{s}m_{t}$) at space constraint interval Σ, where *i* increments *t* from 1 to Σ and *d* decrements *s* from Σ to 1:

$$2^{\Sigma-1} = \sum_{\substack{t_0=1 \\ s_0=\Sigma \\ (i=1,\, d=1)}}^{\substack{t_n=\Sigma \\ s_n=1}} {}^{s_n=s_{n-1}-d}m_{t_n=t_{n-1}+i} \qquad (8.1.1)$$

Example: let Σ=5, then $2^4 = {}^5m_1 + {}^4m_2 + {}^3m_3 + {}^2m_4 + {}^1m_5 = 1+4+6+4+1 = \mathbf{16}$

The finite integral is given as equal to base-2 continuous geometric proportions, where the power of 2 is equal to the spatial interval (Σ) less one (left side of equation 8.1.1), since the singular perspective is 2 to the null power. This base-2 magnitude can also be found by summing differentiated spatial magnitudes associated with translational perspectives (right side of equation 8.1.1). Translational spatial perspective is decremented and temporal differentiation is incremented through the range of state magnitudes for a given spatial constraint interval. Corresponding anti-field integrals may be defined analogously by using temporal constraint intervals (T) instead. It should be evident that equation 8.1.1 relies on the fact that Σ and T are equal in magnitude and opposite in direction. Therefore, only one of the two is required to describe intervals when one uses both increments and decrements.

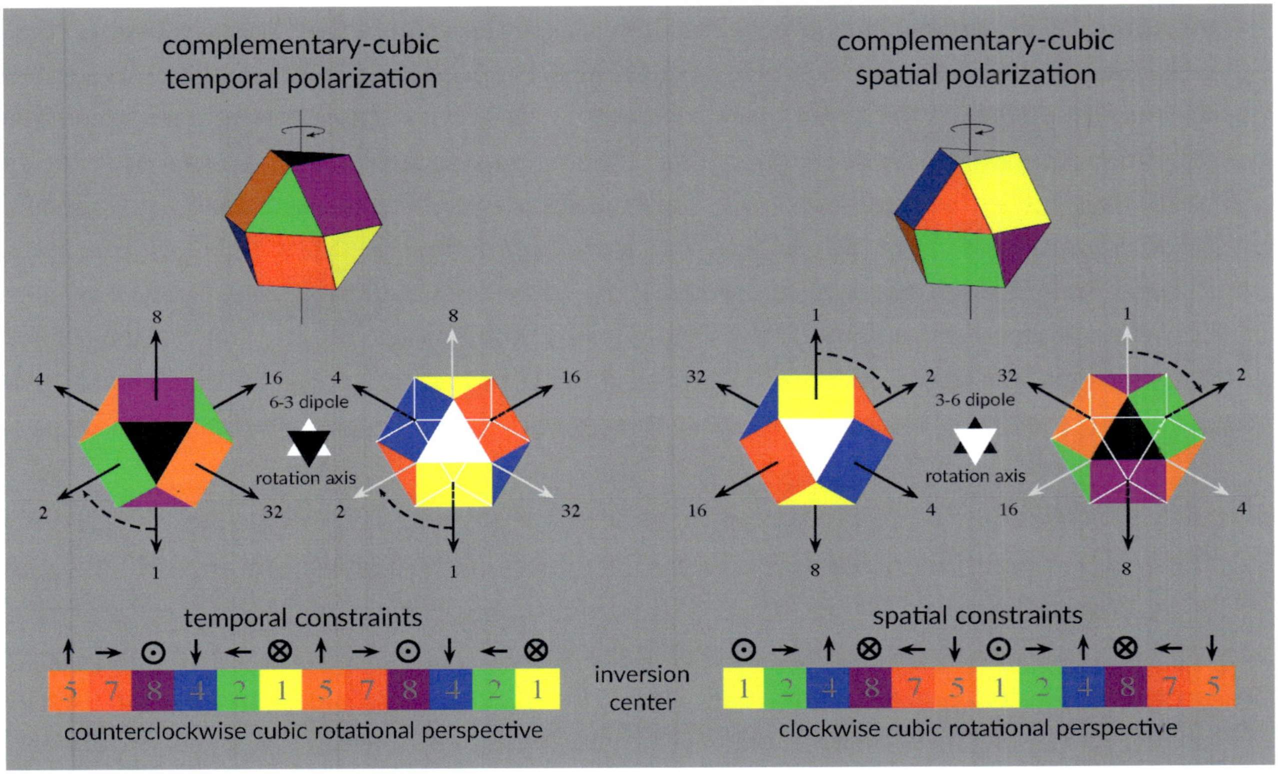

Figure 24. Complementary-cubic spatial and temporal rotational perspectives: Spatial and temporal constraints are shown just above their respective complementary-cubic chromo-numeric polarities. Cubic rotational transformations are shown above their respective 6-fold analog series with dashed arrows indicating the direction of rotation and solid arrows indicating the directions of polarization that correspond to the constraint orientations shown in the lower portion of the figure. To show the polarities of the back faces the front polarities are removed leaving white outlines.

Recall that in the context of Pascal's Triangle base-2 continuous geometric proportion is a mathematical representation of the 6-fold rotational super-symmetry of the complementary color cube, which is shown in Chapter 7 (Figure 22). Figure 24 details the same symmetry for both spatial and temporal rotational perspectives in full cuboctahedral context. The spatial (clockwise) and temporal (counter-clockwise) rotational constraint patterns of Pascal's Triangle are provided along side their complementary-cubic polarity transformations in the figure to illustrate the correspondence between complementary polarities and their relative constraint orientations.

In Chapter 6 I said that the translational perspectives of 3 and 6 were unique relative to other complements since for instance their 3-6 spatial-spatial constraints were anti-parallel and their 3-6 spatial-temporal constraints were parallel, and I indicated that other complements have similar characteristics in a rotational context. In the super-symmetric rotational context the other complements are so constrained: for instance, the constraint of the spatial polarity-1 aligns anti-parallel with the constraint of spatial polarity-8 and parallel with the constraint of temporal polarity-8. This illustrates that the constraints of all cubic spatial-spatial complements are anti-parallel (oppositely constrained) and that the constraints of all cubic spatial-temporal complements are parallel (co-constrained). Given the qualities of complementary triangular dipoles previously noted, shape orientation symmetry suggests that anti-parallel complements symmetrically neutralize one another and that parallel complements do not. For instance, complementary faces of the same cubic super-symmetric perspective are constrained anti-parallel (opposite faces) and sum to 9 or grey, whereas complementary faces of differing super-symmetric perspectives are constrained parallel (same face orientation) and do not sum to 9 or grey since one is temporal (negative) and the other is spatial (positive).[*]

8.1.2 COMPLEMENTARY-CUBOCTAHEDRAL ROTATIONAL TRANSFORMATIONS

The integral proportions of combined spatial/temporal intervals are shown in Figure 25. Here, following from Figure 23, the incremental imposition of spatial constraints is accounted for and spatial/temporal intervals are aligned in rows. When the resulting rows are summed, spatial/temporal intervals form the continuous golden proportion of the "Fibonacci" series.[†] The anti-perspective (temporal/spatial) mirrors the same proportion, but in columns.

Similar to spatial and temporal intervals, finite spatial/temporal and temporal/spatial constraint intervals can be defined by using finite integral notation, as given with equation 8.1.2. Let ${}^{\Sigma/\mathrm{T}}E_{\mathrm{I}}$ signify the finite integral of space/time (Σ/T) at their combined constraint interval (I), where *d* is the decrement of space and *i* the increment of time:

$$ {}^{\Sigma/\mathrm{T}}E_{\mathrm{I}} = \sum_{\substack{s_0=\mathrm{I} \\ t_0=1 \\ (i=1,\ d=2)}}^{\substack{s_n=1\ (\text{odd I}) \\ s_n=2\ (\text{even I}) \\ t_n=(\mathrm{I}+1)/2\ (\text{odd I}) \\ t_n=\mathrm{I}/2\ (\text{even I})}} {}^{s_n=s_{n-1}-d}m_{t_n=t_{n-1}+i} \qquad (8.1.2) $$

Examples:
Let I=7, then ${}^{\Sigma/\mathrm{T}}E_7 = {}^7m_1 + {}^5m_2 + {}^3m_3 + {}^1m_4 = 1+5+6+1 = \mathbf{13}$
Let I=6, then ${}^{\Sigma/\mathrm{T}}E_6 = {}^6m_1 + {}^4m_2 + {}^2m_3 = 1+4+3 = \mathbf{8}$

---

[*]Note that a neutral inversion center separates spatial and temporal perspectives. This fact will become more apparent during the coming analysis of the complementary-cuboctahedral symmetry transformations of continuous golden proportions.

[†]Many internet sources suggest that Pingala referred to the Fibonacci series as Mātrāmeru, though I haven't found any direct citation confirming this. Mātrā means cadence or rhythm.

These integrals correspond with the spatial/temporal constraint intervals shown in Figure 25. Initial boundary conditions are used to fix the beginning of the integral based on a given spatial/temporal constraint interval. The spatial/temporal intervals of Figure 25 are asymmetric with respect to spatial and temporal constraints. This is readily apparent since every other interval (row) lacks a singular temporal state. The reason for this is that the Fibonacci series is formed from the shallow diagonal (2:1) of Pascal's Triangle. Space is decremented by 2 for every time increment of 1 in equation 8.1.2 to form the intervals owing to the 2:1 (spatial:temporal) constraint asymmetry of spatial/temporal intervals. Hence, even and odd intervals must be treated as unique cases.

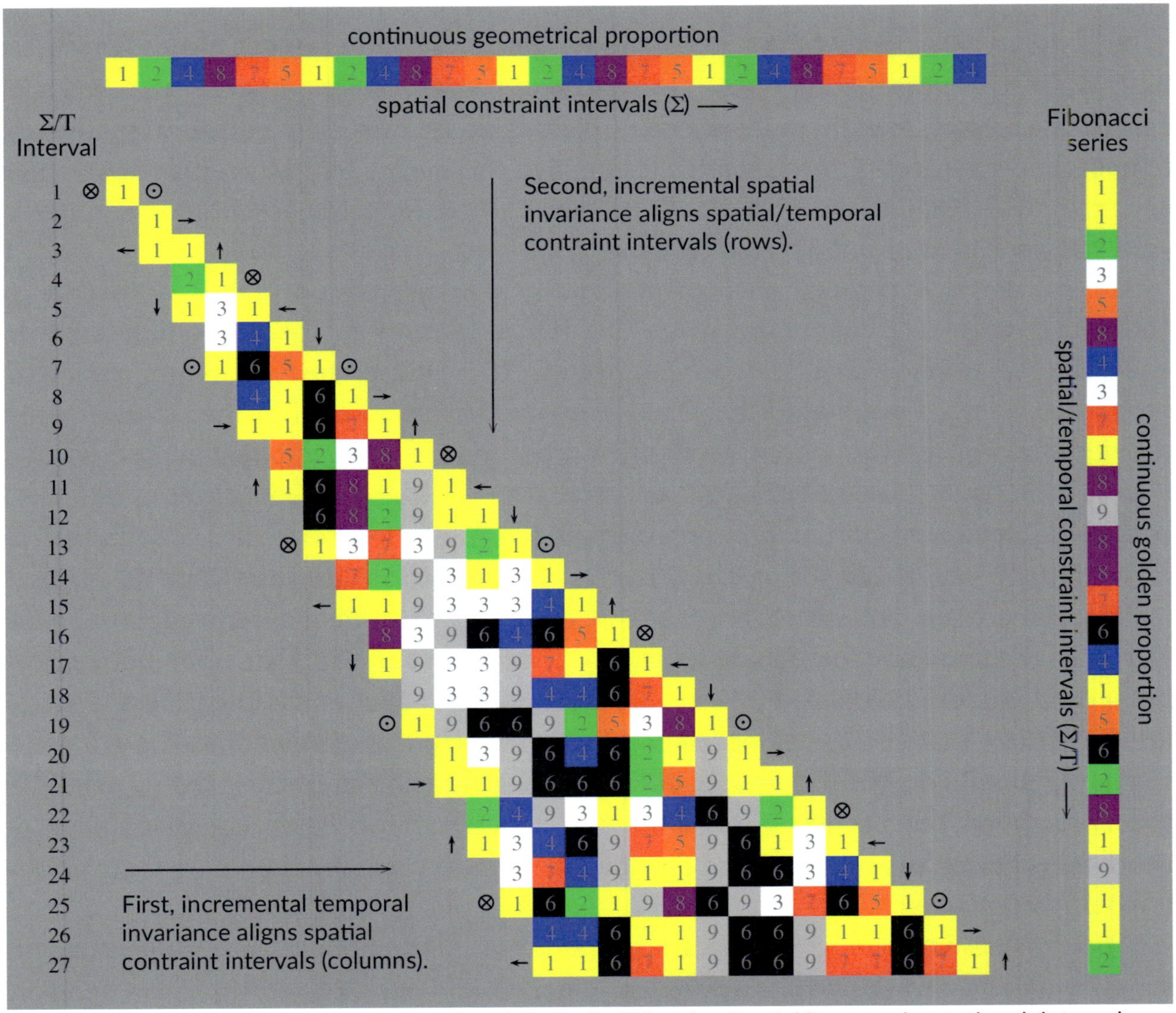

Figure 25. The super-symmetric intervals of Pascal's Triangle: Spatial/temporal rotational intervals are isolated by accounting for incremental imposition of temporal then spatial constraints (step 1 then step 2). These intervals correspond to the continuous golden proportions of the Fibonacci series (shown on the far right in modulus-9 chromo-numeric form). The spatial intervals along the top carry over from Figure 23.

The modulus-9 analysis in Figure 25 shows that the continuous proportion of the Fibonacci series is accompanied by a repeating twenty-four step complementary polarity transformation sequence, where polarity-1 inverts to its complement polarity-8 after a series of twelve transformations and then after an additional twelve numeric transformations returns to its original super-symmetric polarity-1 field state. This suggests that the Fibonacci series is associated with a 1-8 dipole inversion symmetry transformation.

The neutral states at the 12th and 24th spatial/temporal intervals imply that a neutral state exists also at the 0th interval. The neutral state at the 0th interval aligns with the inversion center between spatial and temporal perspectives shown in Figure 24. This initial neutral state corresponds to the zero at the beginning of the Fibonacci series, which can be calculated using equation 4.4.6. The reader will recall that the interval-zero inversion center is associated with the primordial constraint of inversion between spatial and temporal perspectives. From a spatial perspective this means that the center of inversion constrains the point to an introverted perspective that reflects back from the mirror of void anti-parallel to its extroverted temporal antithesis.

In Chapter 4 I showed that continuous golden proportion is a special case of continuous geometric proportion where the future term of such a proportion is always equal to the sum of past and present terms. This special *golden* case of geometric proportion is more symmetric than base-2 geometric proportion. In that chapter I also showed that a given continuous geometric proportion has one exponential base equal to the ratio of any two sequential terms. In contrast the ratios of sequential terms of a given continuous golden proportion only approximate the golden ratio since it is only one of two ratios descriptive of such a continued proportion, the golden ratio and its negative reciprocal (base and anti-base). The fact that the ratios of sequential terms do not equate with a single base like the ordinary geometric proportion suggests that continuous golden proportion is more complex and implies that a purely spatial or temporal perspective is insufficient to characterize its accompanying symmetry transformations. The combined spatial-temporal constraint intervals in Figure 25 affirm this and show that a combined spatial-temporal perspective is required to characterize the Fibonacci series since its continuous golden proportion is the result of the ordered combination of both spatial and temporal constraints.

The constraint patterns of spatial, temporal, spatial/temporal, and temporal/spatial super-symmetric dimensional intervals are detailed in Figure 26. In this figure temporal constraints have been transposed from right-to-left (Figure 24) to left-to-right so that spatial and temporal constraints can be examined side-by-side. Note that the spatial and temporal constraints of base-2 geometric proportions (a.k.a. complementary-cubic proportions) are symmetric with respect to one another. In contrast, the spatial and temporal constraints of the Fibonacci series (a.k.a. complementary-cuboctahedral proportions) are asymmetric (2:1). So the asymmetry of dimensional constraints results in the formation of distinct spatial/temporal and temporal/spatial rotational perspectives derived from mirrored sets of differentiated states. This is in contrast to the spatial and temporal cases of continuous geometric proportion, where the same states are simply observed from oppositely constrained perspectives.

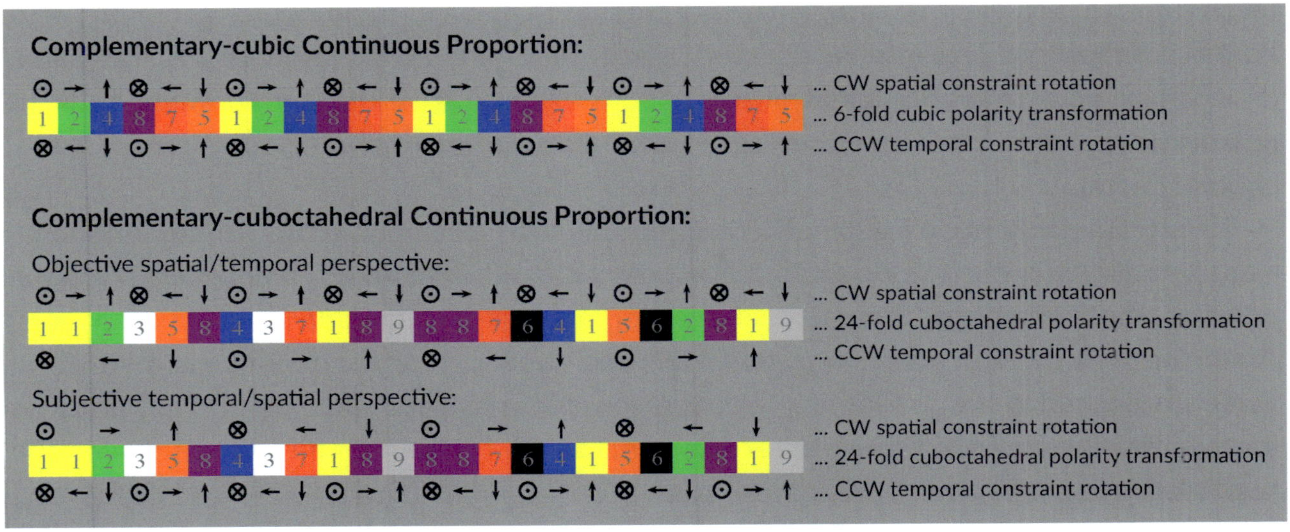

Figure 26. Dimensional constraint rotation and accompanying modulus-9 polarity transformations are shown for spatial and temporal complementary-cubic proportions and for spatial/temporal and temporal/spatial continuous golden proportions (complementary-cuboctahedral). Temporal constraints are transposed from right-to-left (Figure 24) to left-to-right to enable side-by-side comparisons.

In the case of the continuous golden proportion the constraint pattern is 12-fold due to the 2:1 juxtaposition of spatial and temporal constraints, yet the modulus-9 polarity pattern is 24-fold. I've associated this 24-fold symmetry transformation of the Fibonacci series with complementary-cuboctahedral geometry. Now I'll explain why I've come to this conclusion. Notice that the constraints associated with the first polarity-1 (Figure 26, far left) carry over from cubic proportions to cuboctahedral proportions. After this initial state, the constraint patterns diverge and then converge again every 12 steps. This suggests that both initial states share the same cubic symmetry with complementary-cubic proportions (i.e. they still represent cube faces of polarity-1). This follows readily since they all represent aspects of the same originating monad perspective.

Subsequent to twelve symmetry transformations, polarity-8 is constrained the same as the original cube-face of polarity-1. This suggests that this state of polarity-8 is also cubic since it is in a complementary analog position to the initial cube-face polarity-1. Yet recall from Figure 24 that the cubic polarity-8 constrained parallel to the spatial cubic polarity-1 is a temporal cubic polarity. Hence, constraint symmetry suggests that the Fibonacci series is accompanied by transformations between co-constrained 1-8 spatial-temporal dipoles or between co-constrained 1-8 temporal-spatial dipoles.

The preceding analysis suggests that two out of the twenty-four continuous golden proportion symmetry transformation states are cubic, but what about the rest? I think it's reasonable to assume that the symmetry transformations of continuous geometric and golden proportion are inter-connected since they are both ordered by the same dimensional constraints. Polarity transformations alone suggest that the former is cubic (6 polarities) and the latter is either octahedral (8 polarities) or a combination of both, cuboctahedral.

Recall that continuous geometric proportions are accompanied by 6-fold rotational transformations about a 3-6 octahedral dipole axis and that these complementary-cubic polarity transformations alternate between primary and secondary phase hemispheres. Assuming that the characteristic symmetry transformations of continuous geometric and golden proportions are similar, it's logical to predict that continuous golden proportions involve rotation about a dipole axis, and to expect that its polarities alternate between the two hemispheres relative to said dipole axis.

Based on the preceding analysis, the symmetry transformations of the Fibonacci series appear to be rotational transformations about a 1-8 spatial-temporal cubic dipole axis. This implies that the symmetry transformations that accompany the Fibonacci series are a combination of both spatial and temporal complementary-cubic rotational symmetries shown in Figure 24. To confirm this interpretation spatial and temporal complementary cuboctahedrons must be reoriented to the rotational axis under consideration.

To evaluate complementary-cubic transformations we were oriented to 3-6 octahedral dipole axes. To evaluate 1-8 spatial-temporal rotational transformations we must look instead down the 4-fold cubic axes of spatially and temporally polarized complementary cuboctahedrons. Figure 27 shows the spatial 1-8 cubic dipole axis oriented with polarity-1 facing out from the page and shows the temporal 1-8 cubic dipole axis oriented with polarity-1 facing into the page. Both poles under evaluation (spatial polarity-1 and temporal polarity-8) then face out of the page. This will allow us to evaluate the possible chromo-numeric symmetry transformations that correspond to the Fibonacci series polarity sequence.

Looking down the 4-fold axes of both complementary cuboctahedrons (Figure 27) and flipping from top to bottom hemispheres the octahedral green face (2) is below the octahedral blue face (4). Similarly orange (5) is below red (7), and so on. This is true for both spatial and for temporal polarization. By examining the cuboctahedral polarization associated with the spatial cubic perspective under counterclockwise rotation and by alternating hemispheres back-front-back-front, one sees a 4-fold continuous golden proportion transformation sequence of green (2), white (3), orange (5), and violet (8) that matches the beginning of the Fibonacci series shown in bold (0, 1, 1, **2**, **3**, **5**, **8**). In addition, there is a second counterclockwise 4-fold sequence that corresponds to another segment of the 24-fold golden proportion transformation sequence where instead polarities alternate between hemispheres front-back-front-back: red (7), black (6), blue (4), and yellow (1).

Next, examining the clockwise rotation instead, two more 4-fold patterns are apparent. But there's a telling interpretation complication (a mismatch between the physical model and the numeric analysis): where one expects to find a six there is instead a three, and where one expects to find a three there is instead a six. However, remember the unique character of 3-6 dipoles, namely that the translational constraints of 3 and 6 are anti-parallel for either spatial-spatial or temporal-temporal complements and parallel for spatial-temporal or temporal-spatial complements. Uniquely the polarity-3 and polarity-6 triangles that form octahedral faces appear to behave like changelings masquerading as each other (for instance refer to Figure 10 where 3-like triangular planes alternate 3-6-9 and 6-like planes alternate 6-3-9). If you consider that rotation is in the opposite direction relative to the prior 4-fold transformations, it becomes evident that this polarity-3 could very well be a 6-like octahedral plane and polarity-6 could be a 3-like octahedral plane in terms of their respective triangular face orientations. This interpretation is consistent with the nested complementary dipole inversions present in the field that will be discussed in more detail in association with Figure 33.

The same set of polarity transformations are found for temporal cubic polarization as well. Due to the constraint orientations adopted from Figure 26, rotation transformations for spatial and temporal polarization appear to be in the same direction. However from the temporal perspective, which is equivalent to rotating the temporal cubic polarity-1 to the front face, rotational polarity transformations occur in the opposite direction relative to the spatial cubic polarity-1 perspective. I use the spatial orientation as the reference perspective in order to maintain consistency in constraint orientation. This makes the temporal perspective appear backward.

Having catalogued all of the possible 4-fold octahedral transformations in Figure 27 only six of the twenty-four Fibonacci series polarity transformations remain to be accounted for: two neutral states, the polarity states that precede each neutral state, and the states that follow the cubic dipole rotation axis states. Given that both cubic and octahedral states have been identified, it is likely that the states yet to be accounted for are associated with inversion between cubic and octahedral polarity states and with inversion between spatial and temporal rotational perspectives.

My analysis suggests that the states preceding the inversion centers are octahedral since these states are the result of inversion transformations like the octahedral inversion transformations elsewhere in the sequence. The inversion centers are therefore formed by the combination of octahedral poles that are spatial-temporal complements. This implies that the inversion centers are responsible for inversion between octahedral and cubic symmetries as well as between spatial and temporal polarization. This leads to the conclusion that the states that immediately follow cubic poles are octahedral poles that have opposite spatial-temporal polarization from the cubic pole, since they are formed when a given cubic pole is summed with the inversion center that precedes it and since this octahedral face is in the correct rotational position to participate in the transformations that follow.

The reader should be aware that they might find it difficult to follow and confirm the preceding line of reasoning for themselves with the 2D graphics alone (i.e. without physical models in hand). Such models may be constructed using the templates provided in Appendix B. My basis for concluding that the Fibonacci series is accompanied by cuboctahedral symmetry transformations stems from observations of the 4-fold rotational symmetry transformations shown in Figures 27, 29, and 31 coupled with the readily apparent complementary numeric transformations shown in Figures 25 and 26. The preceding is my best attempt at connecting the dots between these 4-fold segments, but there is presently no objective means of conclusively validating these portions of the transformation sequence. There are other possible interpretations of a few of the transformation elements that I've intuited based on my own fallible logic. Therefore, I encourage investigation and discourse on the subject so that a consensus may be reached.

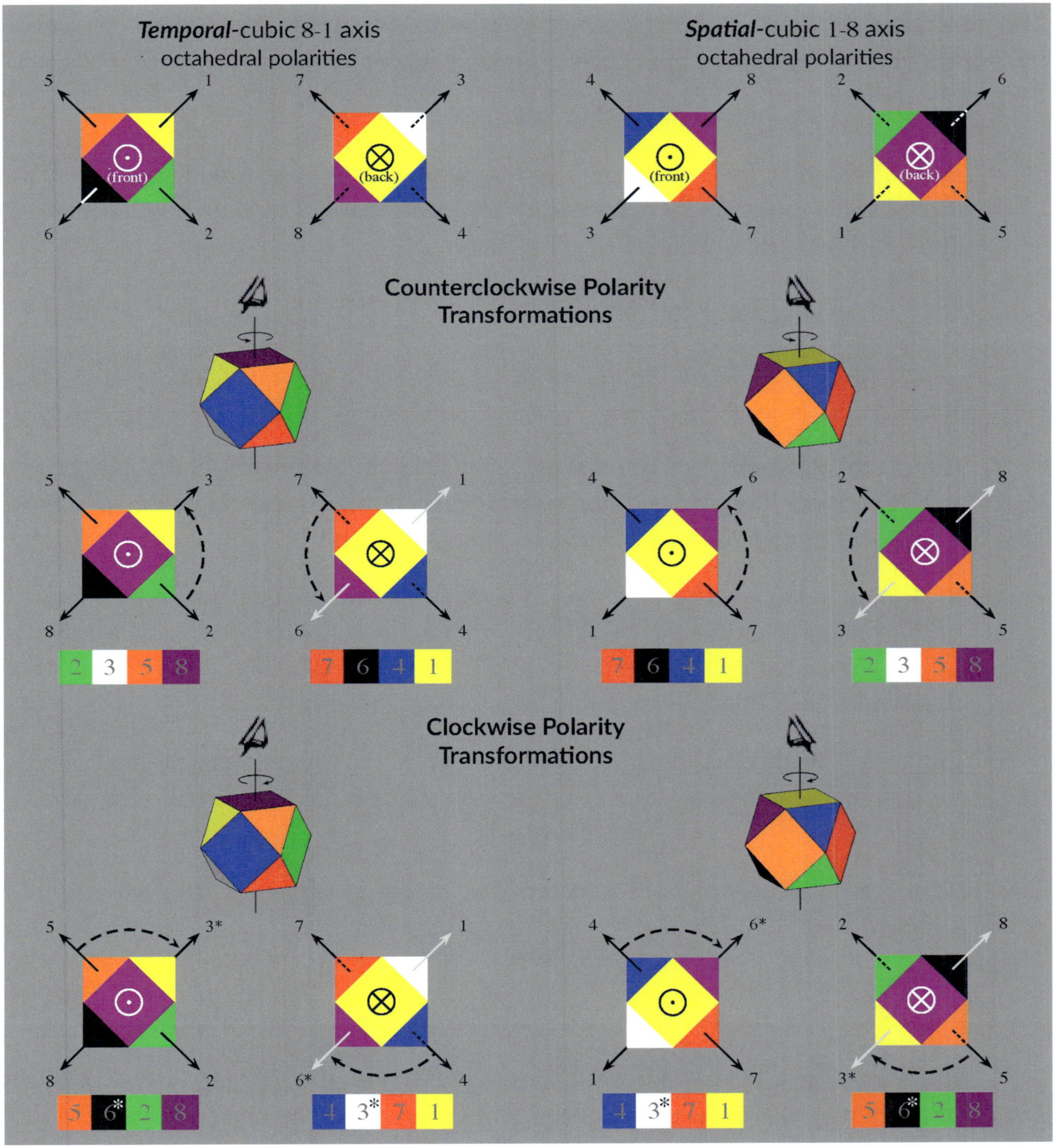

Figure 27. Spatial and temporal cubic 1-8 dipole rotation axes: Polarity analysis reveals 4-fold octahedral Fibonacci field segments. Asterisks indicate that the number 3 is paired with octahedral face polarity-6 and vice versa. Front views are indicated by circle-dot and back by circle-x. Arced dashed arrows represent rotation direction. Straight arrows represent polarization direction. Back views are shown from the perspective of the front so front arrows are shaded white and back arrows have an initial dash to indicate they are directed from the backside of the figure at an angle into the plane of the page.

The following sequence summarizes my interpretation of the cuboctahedral spatial-temporal inversion sequence of the cubic 1-8 dipole field based on the preceding analysis:

- The sequence begins at the spatial cubic pole (polarity-1).*
- Then, given that the future state is the sum of the present and past state, the spatial cubic pole is transformed to the temporal octahedral pole (polarity-1) through summation with the inversion center.
- Next the temporal octahedral pole (polarity-1) transforms to spatial octahedral polarity-2 through counterclockwise rotation from the front hemisphere to the back hemisphere followed by three spatial cubic 90° counterclockwise rotations to octahedral polarity-8.†
- The spatial octahedral polarity-8 then combines with the polarity-5 that preceded it and both orientation and rotation reverse causing the spatial octahedral polarity-5 to invert to its complement: temporal octahedral polarity-4.
- Next temporal octahedral polarity-4 is summed with spatial polarity-8 and, through a 90° temporal clockwise rotation, is transformed to the temporal octahedral polarity-6 represented numerically as 3. This is followed by two more 90° temporal clockwise transformations to temporal octahedral polarity-1.‡
- Temporal octahedral polarity-1 then inverts to its co-constrained spatial complement, octahedral polarity-8, through summation with temporal octahedral polarity-7.
- Next the temporal-spatial 1-8 octahedral dipole combines to form the neutral inversion center.
- The final step in the cubic dipole inversion occurs when spatial octahedral polarity-8 combines with the inversion center and transforms to temporal cubic polarity-8.
- The following twelve transformations complement the former twelve, thereby inverting temporal cubic polarity-8 back to spatial cubic polarity-1.

---

*Note that the continuous golden proportion can also begin with temporal cubic polarity-1, as in Figure 26; this is the chiral perspective of the temporal/spatial anti-field.

†Spatial octahedral polarity-8 is the center of the twelve-step inversion from spatial to temporal cubic poles.

‡Note that temporal-clockwise rotation is equivalent to spatial-counterclockwise rotation when in the temporal cubic orientation where polarity-1 is the front-face as opposed to the back face as shown in Figures 27 and 28.

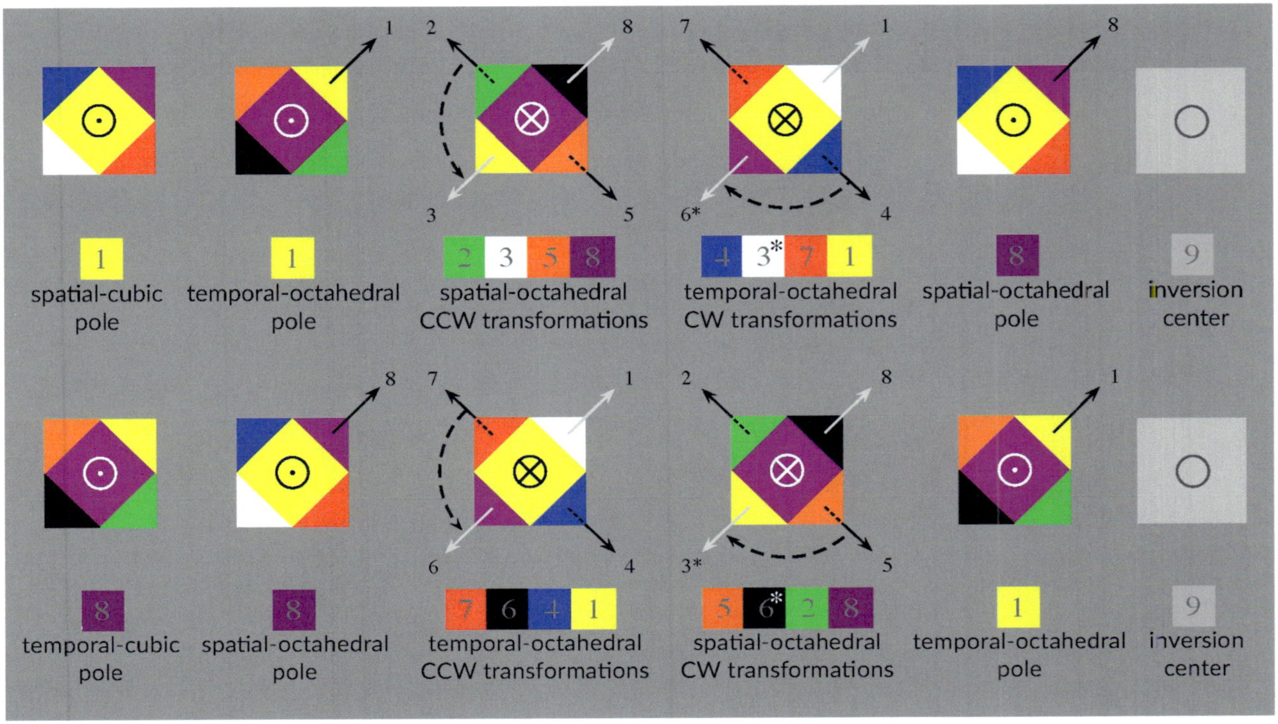

Figure 28. Fibonacci series: Modulus-9 analysis shows that the 24-fold complementary numeric transformations of the continuous golden proportion are accompanied by super-symmetric cuboctahedral rotational transformations about the 1-8 spatial-temporal cubic dipole axis.

If the "Fibonacci" series represents either a spatial-temporal or temporal-spatial 1-8 cubic dipole inversion field, it follows that other continuous golden proportions such as the "Lucas" series might be similarly characterized. The Lucas series begins 2, 1, 3, 4, 7, 11, and so on. Complementary polarity transformations suggest that the Lucas series is a 7-2 cuboctahedral dipole inversion field. So next we reorient the spatial chromo-numeric super-symmetry to the 7-2 (front-back) cubic dipole axis orientation and reorient the temporal to the 2-7 (front-back) cubic dipole axis orientation.

Analysis of the 7-2 cubic dipole rotation axes reveals again that there are eight 4-fold octahedral golden proportion transformation segments that are divided between spatial and temporal cubic perspectives and between clockwise and counterclockwise rotation (Figure 29). Combining these segments, as before, agrees with the 24-fold 7-2 spatial-temporal cubic dipole inversion field that accompanies the Lucas series (Figure 30). In this analysis the dipole field is oriented spatial pole to temporal pole (the same as Figure 28).

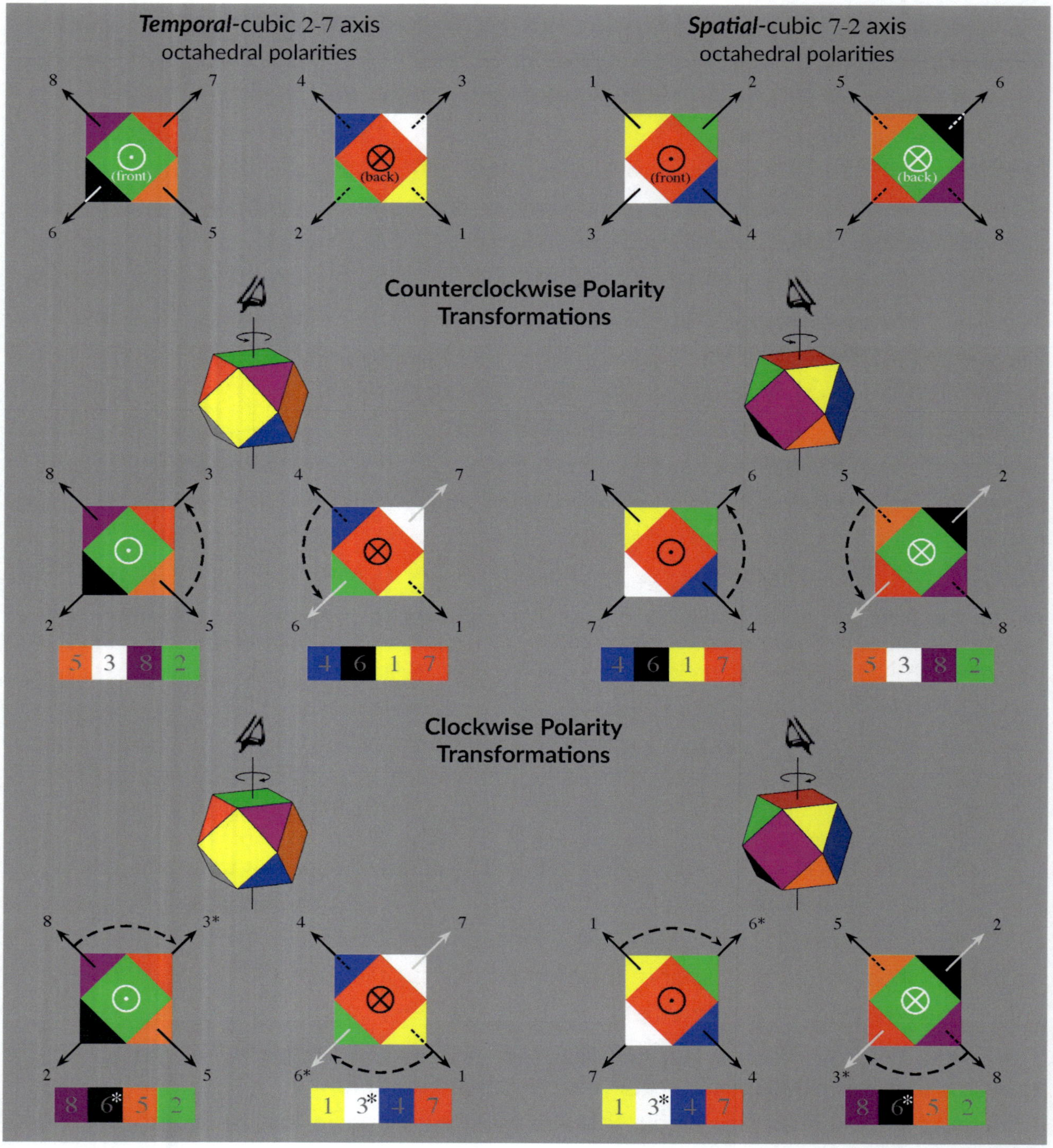

Figure 29. Spatial and temporal cubic 7-2 dipole rotation axes: Polarity analysis reveals 4-fold golden proportion field segments. Asterisks indicate that the number 3 is paired with octahedral face polarity-6 and vice versa. Front views are indicated by circle-dot and back by circle-x. Arced dashed arrows represent rotation direction. Straight arrows represent polarization direction. Back views are shown from the perspective of the front so front arrows are shaded white and back arrows have an initial dash to indicate they are directed from the backside of the figure at an angle into the plane of the page.

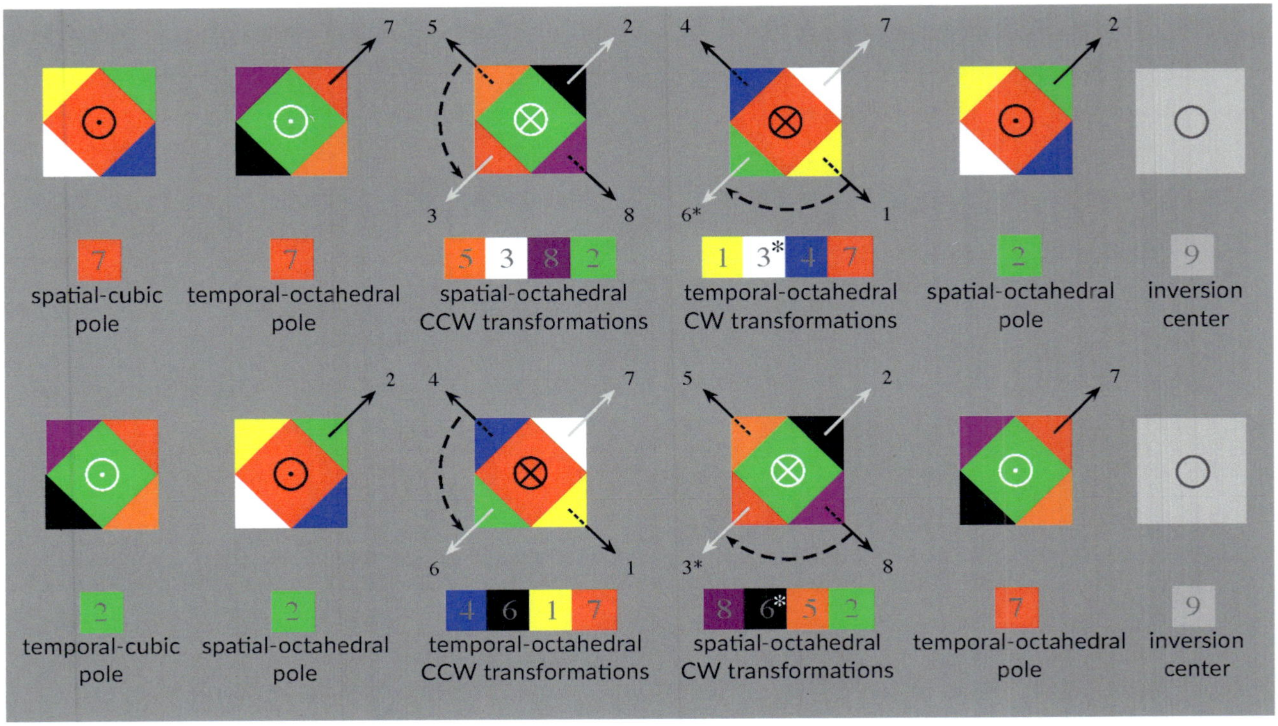

Figure 30. Polarity transformation analysis shows that rotation about the 7-2 spatial-temporal cubic dipole axis supports 24-fold complementary cuboctahedral geometric transformations associated with continuous golden proportions. The "Lucas" series (2, 1, 3, 4, 7, 11, and so on) begins at the sixth step of the 7-2 cubic dipole inversion sequence.

The sixth step of the 7-2 cubic dipole inversion sequence shown in Figure 30 corresponds with the beginning of the Lucas series, which indicates that the series initiates with the spatial octahedral polarity-2. Note that this polarity state marks the center of the twelve-step spatial-to-temporal cubic dipole inversion sequence. The initial polarity-2 of the Lucas series corresponds with the initial neutral polarity of the Fibonacci series – the neutral inversion center between the field and the anti-field.

For completeness I have also included an analysis of 4-5 cubic dipole rotation axes. The analysis indicates that there are also eight 4-fold octahedral golden proportion transformation segments, which are divided equally between spatial and temporal cubic perspectives as well as between clockwise and counterclockwise rotation (Figure 31). As before, combining these segments produces the 24-fold analog transformations associated with the either the 4-5 spatial-temporal or temporal-spatial cubic dipole rotation axes. Spatial-temporal transformations are shown in Figure 32. The dipole field is oriented spatial pole to temporal pole (the same way as the prior two cases). Although only spatial-temporal dipole inversions are shown in Figures 28, 30, and 32 equivalent temporal-spatial fields are also supported by the chirality of Pascal's Triangle.

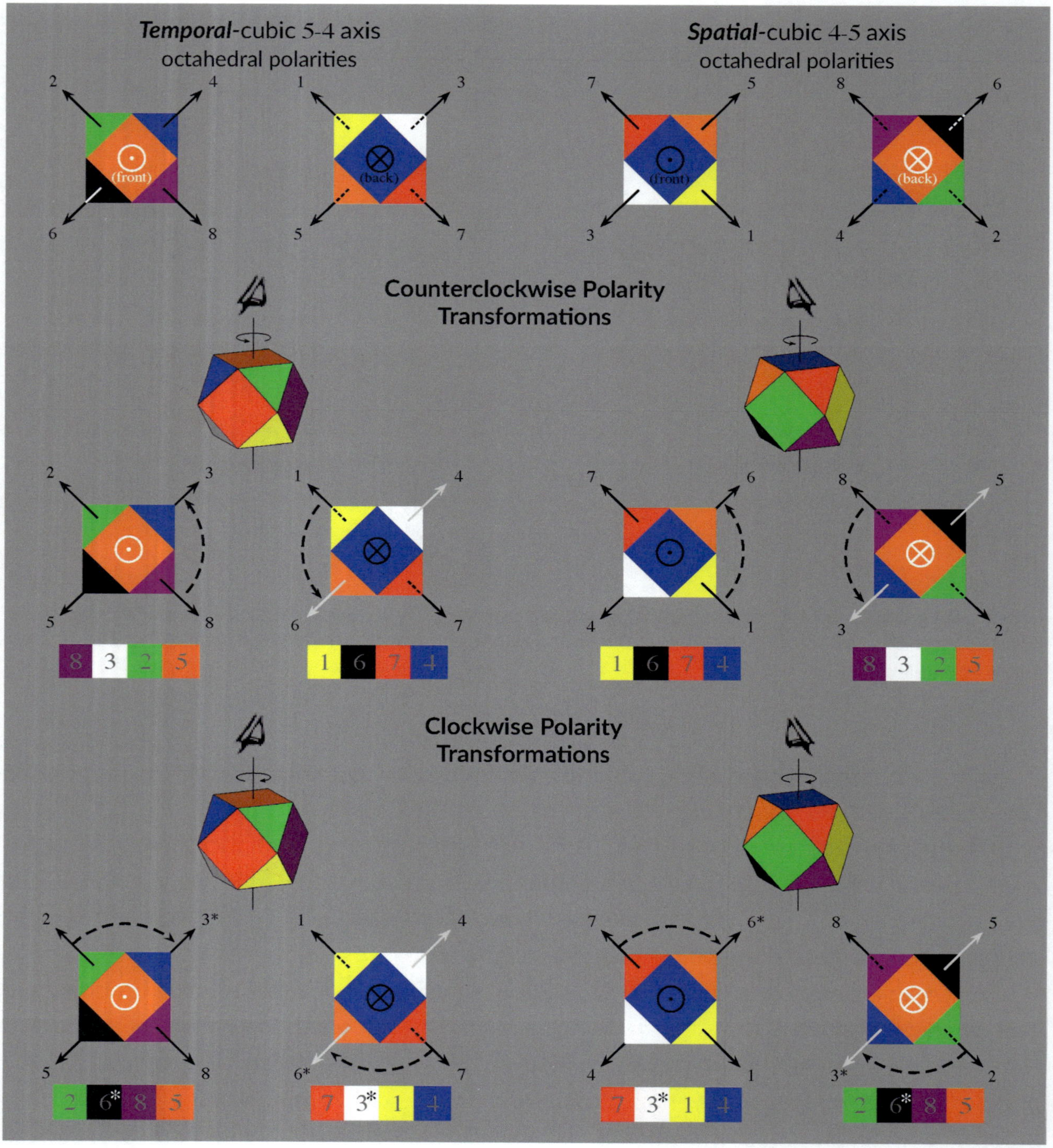

Figure 31. Spatial and temporal cubic 4-5 dipole rotation axes: Polarity analysis reveals 4-fold golden proportion field segments. Asterisks indicate that the number 3 is paired with octahedral face polarity-6 and vice versa. Front views are indicated by circle-dot and back by circle-x. Arced dashed arrows represent rotation direction. Straight arrows represent polarization direction. Back views are shown from the perspective of the front so front arrows are shaded white and back arrows have an initial dash to indicate they are directed from the backside of the figure at an angle into the plane of the page.

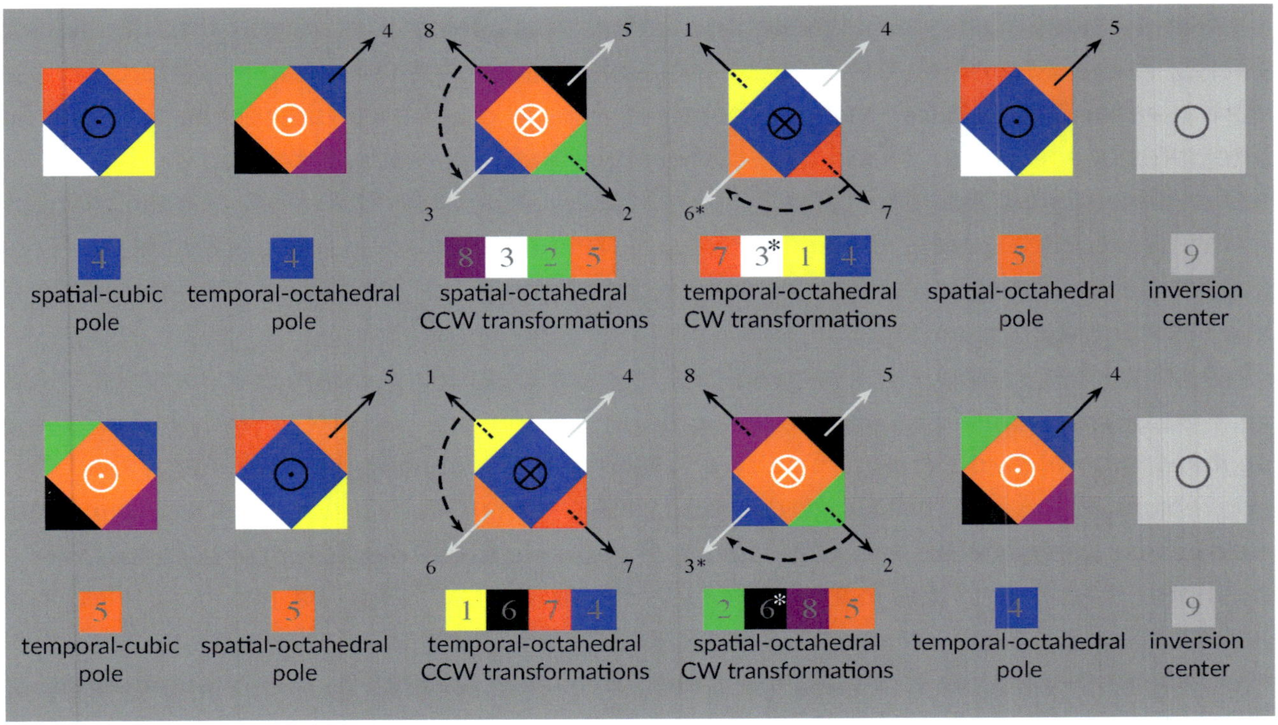

Figure 32. Polarity transformation analysis shows that rotation about the 4-5 spatial-temporal cubic dipole axis supports 24-fold complementary cuboctahedral geometric transformations associated with continuous golden proportions.

8.2 Creative-Force and Receptive-Energy

Until now I've focused on the symmetrical differentiation of energy states and the super-symmetry thereof without considering why energy is impelled to differentiate at all. What drives energy from its singular state, and what caused it to be in that state in the first place? And, if energy is receiving and responding to some impulse, what could energy be symmetrically subordinate to that would create such an initiating impulse and cause a differentiation response?

In classical Newtonian physics the change in kinetic energy of a given mass is equal to the net force imparted to the mass over a given distance; when the force is constant this change in kinetic energy is simply the force multiplied by the distance over which it is applied. This interpretation works for velocities that are small compared to the speed of light. At velocities close to the speed of light the theory of relativity augments the classical model by accounting for the increase in mass of a given object as it approaches the speed of light. Both of these models involve mass since they are developed from our matter-centric perspective; even so they are informative when considering a dimensional perspective where no mass exists yet, a pure energy perspective.

Observations from our matter-centric dimensional perspective suggest that the field of energy is the integral of the field of force or alternately that the field of force is the derivative of the field of energy. Therefore energy can be considered as a field that integrates or absorbs force and thereby becomes more energetic. Force can be thought of as a symmetry field that *initiates* the differentiation *response* of the energy symmetry field. In this sense force *creates* and energy *receives*. And as the former governs the latter, force appears to be senior to energy. In other words force is the initiating cause that governs the differentiation of energy.

As discussed previously, the complementary-cuboctahedral symmetry transformations that accompany the Fibonacci continuous proportion are consistent with either a 1-8 spatial-temporal dipole energy field or a 1-8 temporal-spatial dipole anti-energy field. It is convenient at this juncture to adopt a descriptive naming convention for these symmetry fields since *complementary-cuboctahedral 1-8 spatial-temporal dipole energy field* is a bit of a mouthful. I've named the complementary dipole energy field *The Receptive Field* and named the dipole anti-energy field *The Anti-Receptive Field* since I believe the notion of receiving or responding to an initiating force of symmetry captures the inherent nature of energy.

> Perfect is Responding's [*The Receptive's*] greatness; it brings birth to all beings and accepts the source from Heaven. Responding in its richness sustains all beings; its virtue is in harmony without limit. Its capacity is wide, its brightness is great. Through it, all beings attain their full development.
>
> Confucius [58]

Recall that all forms of continuous proportion are preserved through the continued maintenance of a fixed relationship between three consecutive terms. The fixed relationship for the golden proportion of the energy field is that the present term always equals the difference between future and past terms. This fixed relationship constitutes a symmetric impetus or force of symmetry acting on the present interval by past and future intervals. Past, present, and future energy intervals can be described by two present intervals: *present energy* and *present force*, where present force is the sum of past and future energy intervals. Notice the distinction between the continuous golden proportions of energy and force as defined...

$$\text{ENERGY: } {}^{\Sigma/\mathrm{T}}E_{\mathrm{I}} = {}^{\Sigma/\mathrm{T}}E_{\mathrm{I}+1} - {}^{\Sigma/\mathrm{T}}E_{\mathrm{I}-1} = \frac{\varphi_+^{\mathrm{I}} - \varphi_-^{\mathrm{I}}}{\varphi_+ - \varphi_-}$$

$$\text{FORCE: } {}^{\Sigma/\mathrm{T}}F_{\mathrm{I}} = {}^{\Sigma/\mathrm{T}}E_{\mathrm{I}+1} + {}^{\Sigma/\mathrm{T}}E_{\mathrm{I}-1} = \frac{\varphi_+^{\mathrm{I}} + \varphi_-^{\mathrm{I}}}{\varphi_+ + \varphi_-}$$

The present super-symmetric energy interval is equal to the difference between future and past intervals and it is also described by a difference relationship between the spatial base and temporal anti-base of continuous golden proportion – the difference between future and past and the difference between objective-spatial and subjective-temporal dimensional symmetry elements. In contrast, the present super-symmetric force interval is equal to the sum of future and past energy intervals and it is described by an additive relationship between the spatial base and temporal anti-base of continuous golden proportion – the sum of future and past and the sum of objective-spatial and subjective-temporal dimensional symmetry elements. This provides a vision of the fundamental nature of energy as the antagonism between spatial and temporal symmetry elements (their difference) and a vision of the fundamental nature of force as the cooperation between spatial and temporal symmetry elements (their sum).

It is as if the past were pushing and future were pulling and together they were collaborating to create the force of symmetry that impels the differentiation of energy states in accordance with energy's accompanying complementary-cuboctahedral dipole transformations. The continuous golden proportion that is the sum of past and future Fibonacci series intervals is the Lucas series. As discussed previously, the Lucas series is accompanied by 7-2 dipole complementary-cuboctahedral rotational symmetry transformations. I have named this complementary-cuboctahedral field of force *The Creative Field*, since I believe the *initiation* of a *response* in the energy field *created* by the *receipt* of a symmetrical force accurately depicts the inter-relationship between the two fields under consideration.

> Vast indeed is the greatness of the Initiating [*The Creative*]. It is the source of all beings and regulates all creations under Heaven. ...The way of the Initiating is change and transformation so that each being obtains its true nature and destiny and the union of great harmony is preserved.
>
> Confucius [59]

8.2.1 RECEPTIVE AND CREATIVE FIELD EQUATIONS

In addition to the finite integrals of super-positioned states, the symmetry states of The Receptive Field may also be described as the modulus-9 symmetry transformations that accompany the exponential power series of spatial base and temporal anti-base. This form of golden proportion is equal to the difference between equal powers of base and anti-base divided by their difference (square root of five). As discussed in Chapter 4, the equation for the Fibonacci series is known to science in a form that generally discounts the significance of the negative reciprocal of the golden ratio (the anti-base or subjective temporal root of the continuous golden proportion), and instead the objective spatial root is manipulated to appear temporal. The problem is that in so doing the interplay of the two core symmetry elements is obscured. It is this interplay that suggests a dynamic merger of spatial and temporal symmetry elements that co-differentiate filling all energy states like a multidimensional torus-like waveform.

To place the Fibonacci series in this new context I've rewritten equation 4.4.6 with double vertical brackets indicative of reduction to mod 9 polarities, and I've incorporated the notation introduced with equation 8.1.2 to indicate the super-symmetric spatial/temporal constraint intervals of Pascal's Triangle. I refer to it as *The Receptive Field Equation:**

$$\left\| {}^{\Sigma/\mathrm{T}}E_{\mathrm{I}} \right\| = \left\| \frac{\varphi_{+}^{\mathrm{I}} - \varphi_{-}^{\mathrm{I}}}{\varphi_{+} - \varphi_{-}} \right\| \qquad (8.2.1)$$

Using the same notation, the associated continuous golden proportion equality is rewritten as follows:

$${}^{\Sigma/\mathrm{T}}E_{\mathrm{I}} = {}^{\Sigma/\mathrm{T}}E_{\mathrm{I}+1} - {}^{\Sigma/\mathrm{T}}E_{\mathrm{I}-1} \qquad (8.2.2)$$

Example, interval 0 to 12:

$$0, 1, 1, 2, 3, 5, 8, 13, 21, 34, 55, 89, 144, \ldots, {}^{\Sigma/\mathrm{T}}E_{\mathrm{I}-1}, {}^{\Sigma/\mathrm{T}}E_{\mathrm{I}}, {}^{\Sigma/\mathrm{T}}E_{\mathrm{I}+1}$$

Again using the same notation, I've modified equation 4.4.7 to reflect the new learning regarding spatial/temporal intervals. I refer to it as *The Creative Field Equation*† (8.2.3). It defines the series of modulus-9 symmetry transformations whose continuous golden proportion is described by the exponential power series of the sums of spatial base and temporal anti-base, where consecutive powers equate with consecutive spatial/temporal intervals. Again a similar equation to this is known to science, but generally without the use of the subjective root, which is critical to understanding the core nature of the field:

---

*The Anti-Receptive Field version of Equation 8.2.1 reverses terms so that the anti-base term is subtracted from the base term instead of base minus anti-base. It generates the same continuous proportion since a negative number divided by a negative number is positive. This obscures the fact that the geometry of its continuous proportion is the negative reciprocal of The Receptive Field.

†Equation 8.2.3 can be written with terms reversed, however by the same logic the continuous proportion remains spatially/temporally oriented since there is no sign reversal of numerator and denominator.

$$\left\| {}^{\Sigma/T}F_{I} \right\| = \left\| \frac{\varphi_{+}^{I} + \varphi_{-}^{I}}{\varphi_{+} + \varphi_{-}} \right\| \tag{8.2.3}$$

Similarly, the associated continuous golden proportion equality can be rewritten as follows:

$${}^{\Sigma/T}F_{I} = {}^{\Sigma/T}F_{I+1} - {}^{\Sigma/T}F_{I-1} \tag{8.2.4}$$

Example, interval 0 to 12:

$$2, 1, 3, 4, 7, 11, 18, 29, 47, 76, 123, 199, 322, \ldots, {}^{\Sigma/T}F_{I-1}, {}^{\Sigma/T}F_{I}, {}^{\Sigma/T}F_{I+1}$$

8.2.2 CREATIVE AND RECEPTIVE FIELD RELATIONSHIPS

As discussed, the present term of The Creative Field is equal to the sum of the past and future terms of The Receptive Field. I've named this relationship the *Space/Time Impulse Field*, since I envision *creative*-force as the combined push-pull (past-future) of the symmetry of its continuous golden proportion, which constrains the present *receptive*-energy state. This characteristic of force symmetry is described with equation 8.2.5:

$${}^{\Sigma/T}F_{I} = {}^{\Sigma/T}E_{I-1} + {}^{\Sigma/T}E_{I+1} \tag{8.2.5}$$

Examples at spatial/temporal intervals 5 and 6:

$${}^{\Sigma/T}F_{5} = {}^{\Sigma/T}E_{4} + {}^{\Sigma/T}E_{6} \Rightarrow \quad 7 = 2 + 5$$

$${}^{\Sigma/T}F_{6} = {}^{\Sigma/T}E_{5} + {}^{\Sigma/T}E_{7} \Rightarrow \quad 11 = 3 + 8$$

Numeric analysis reveals that force predicts the rate of differentiation of energy states. For example, compare spatial/temporal intervals 5 and 10 (2x interval-5) where energy interval-5 equals 5 and interval-10 equals 55 (5x11). Notice that the force term at interval-5 equals 11. So the integral magnitude of the dimensional-differentiation of energy symmetry states (5) multiplied by the integral magnitude of force (11) at a given interval is equal to the magnitude of energy at two times the given interval (55). In general then, the product of the present force and energy field interval equals the magnitude of energy differentiation at twice the present interval, or:

$$ {}^{\Sigma\!/\mathrm{T}}E_{2\mathrm{I}} = {}^{\Sigma\!/\mathrm{T}}F_{\mathrm{I}}\,{}^{\Sigma\!/\mathrm{T}}E_{\mathrm{I}} \tag{8.2.6} $$

Example for spatial/temporal interval 5:

$$ {}^{\Sigma\!/\mathrm{T}}E_{\mathrm{I}} :\ 0, 1, 1, 2, 3, \mathbf{5}, 8, 13, 21, 34, \mathbf{55}, \ldots $$

$$ {}^{\Sigma\!/\mathrm{T}}F_{\mathrm{I}} :\ 2, 1, 3, 4, 7, \mathbf{11}, 18, 29, 47, 76, \ldots $$

$$ {}^{\Sigma\!/\mathrm{T}}E_5 = 5 \text{ and } {}^{\Sigma\!/\mathrm{T}}F_5 = 11 \Rightarrow {}^{\Sigma\!/\mathrm{T}}E_{10} = {}^{\Sigma\!/\mathrm{T}}F_5\,{}^{\Sigma\!/\mathrm{T}}E_5 = 11 \otimes 5 = 55 $$

As a result every other Fibonacci series term is the product of prior Fibonacci and Lucas series interval pairs: 0, 1, 3, 8, 21, 55, 144, 337, ... Compare these numbers to the symmetry analysis of Figure 33 and you will find a direct correlation to the interleaving 4-fold transformations contained therein and to their accompanying dimensional constraint pattern. This interdependence implies that the present force term forecasts the rate of growth of energy differentiation. This rate relationship therefore suggests the Lucas and Fibonacci series have some complex form of derivative-integral relationship like force and energy do from our matter-centric Newtonian perspectives.

Recall that, as the nexus of all perspectives, the point represents the geometric origin of both space and time. Yet without the inverting even mirror symmetry caused by the void that surrounds it, the point could not differentiate to fill spatial/temporal and temporal/spatial dimensions with continuous golden proportion. These two perspectives of the singular point arise from the constraint of inversion caused by the mirroring void that surrounds their common singular origin. It follows that the force of symmetry that causes the differentiation of the point is associated with the influence of the surrounding void. This leads me to consider the possibility that the geometric relationship between void and point and the geometric relationship between 7-2 and 1-8 cuboctahedral dipole fields might be related.

The interdependent nature of receptive and creative dipole fields is illustrated in Figure 33. The interconnected symmetry transformations of these two cuboctahedral dipole fields show how the continuous force proportion coordinates the symmetry transformations of the continuous energy proportion. Notice that the initial neutral inversion center of the 1-8 cuboctahedral dipole field of energy coincides with the number 0 and its constraining continuous equal proportion – the even void that internally reflects the point's symmetry upon itself to create virtual spatial-temporal dimensions of the energy field. At this initial zero-power energy interval the difference between spatial base and temporal anti-base is zero, but the force on this inversion center is the even 2, the zero-power sum of spatial base and temporal anti-base. This is the polarity-2 center of an octahedral dipole force inversion field highlighted with nested arcing arrows in Figure 33. The field of force highlights the even mirror symmetry of the void – the neutral inversion center between energy's spatial/temporal and anti-energy's temporal/spatial continuous golden proportions.

Figure 33. Interdependent 1-8 and 7-2 cuboctahedral dipole inversion fields: The neutral inversion centers of the 1-8 dipole field coincide with the octahedral centers of inversion between cubic poles of the 7-2 dipole field, as distinguished with dashed lines. A 4-fold pattern of polarity inversion transformations is highlighted using arced lines with arrows. Nested dipole inversions are shown with concentric arced lines.

A 4-fold pattern of polarity inversion transformations is also highlighted in Figure 33 with transformation arrows. Within the 1-8 dipole field, 4-fold tertiary phase polarity transformations are centered between 4-fold 1-8 octahedral inversion transformations. Within the 7-2 dipole field, 4-fold tertiary phase polarity transformations are centered between 4-fold 7-2 octahedral polarity inversion transformations. The juxtaposition of the two fields aligns tertiary phase polarities of the 1-8 cuboctahedral field with 7-2 octahedral polarities of the 7-2 cuboctahedral field. Likewise tertiary phase polarities of the 7-2 field are aligned with 1-8 octahedral polarities of the 1-8 field. These 4-fold transformations correspond to the power structure within the Lucas series' 7-2 cuboctahedral dipole field described by the following relationships:

Even intervals create a 4-fold power structure between complements 2 and 7 per equation 8.2.7. **Since for all even intervals...**

$$ {}^{S/T}F_{2\mathrm{I}} = {}^{S/T}F_{\mathrm{I}}^{2} - 2 \qquad (8.2.7) $$

$$ {}^{S/T}F_{0} = 2 \ \Rightarrow\ {}^{S/T}F_{0} = {}^{S/T}F_{0}^{\,2} - 2 = 2^{2} - 2 = \mathbf{2} $$

$$ {}^{S/T}F_{2} = 3 \ \Rightarrow\ {}^{S/T}F_{4} = {}^{S/T}F_{2}^{\,2} - 2 = 3^{2} - 2 = \mathbf{7} $$

$$ {}^{S/T}F_{4} = 7 \ \Rightarrow\ {}^{S/T}F_{8} = {}^{S/T}F_{4}^{\,2} - 2 = 7^{2} - 2 = \mathbf{47} $$

$$ {}^{S/T}F_{\mathrm{I}} : \mathbf{2}, 1, 3, 4, \mathbf{7}, 11, 18, 29, \mathbf{47}, 76, 123, 199, \mathbf{322}, 521, 843\ldots $$

Odd intervals create a 4-fold power structure between tertiary phase polarities per equation 8.2.8. **Since for all odd intervals...**

$$ {}^{S/T}F_{2\mathrm{I}} = {}^{S/T}F_{\mathrm{I}}^{2} + 2 \qquad (8.2.8) $$

$$ {}^{S/T}F_{1} = 1 \ \Rightarrow\ {}^{S/T}F_{2} = {}^{S/T}F_{1}^{\,2} + 2 = 1^{2} + 2 = \mathbf{3} $$

$$ {}^{S/T}F_{3} = 4 \ \Rightarrow\ {}^{S/T}F_{6} = {}^{S/T}F_{3}^{\,2} + 2 = 4^{2} + 2 = \mathbf{18} $$

$$ {}^{S/T}F_{5} = 11 \ \Rightarrow\ {}^{S/T}F_{10} = {}^{S/T}F_{5}^{\,2} + 2 = 11^{2} + 2 = \mathbf{123} $$

$$ {}^{S/T}F_{\mathrm{I}} : 2, 1, \mathbf{3}, 4, 7, 11, \mathbf{18}, 29, 47, 76, \mathbf{123}, 199, 322, 521, \mathbf{843}\ldots $$

The nested complementary octahedral inversion transformations within each 12-step cubic dipole inversion validate the prior proposal that the second 3 within the first 12-tep transformation is 6-like and that the second 6 within the second 12-step transformation is 3-like since in each case the inversions are between complements. This serves as further validation of the complementary cuboctahedral symmetry first proposed in Chapter 5.

Another noteworthy contrast between these two fields is that the Ith root of the Ith term of the Fibonacci series approaches the golden ratio as a limit, whereas the Ith root of the Ith term of the Lucas series rapidly converges on the ideal ratio. One can see why this is by inspecting their respective field equations. In the case of force, as the spatial/temporal interval increases the magnitude of the Ith power of temporal anti-base approaches zero, hence the Ith root of the Lucas series rapidly converges on the value of spatial base ($\boldsymbol{\varphi_{+}}$). In the case of energy, the $\sqrt{5}$ divisor causes I^{th} root of the Fibonacci series to asymptotically approach the value of the spatial base. Both cases are contrasted graphically in Figure 34.

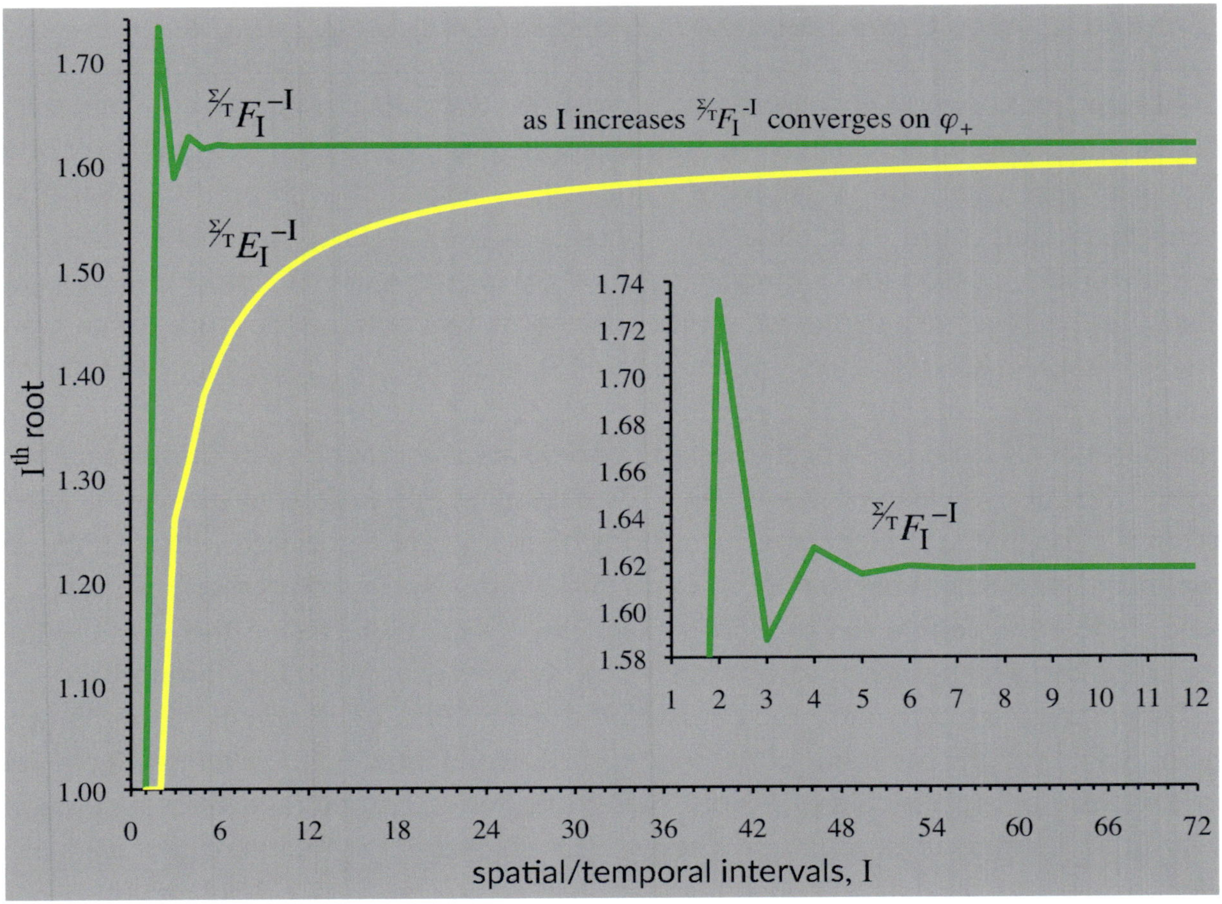

Figure 34. The plot of the Ith roots of Lucas and Fibonacci series terms demonstrates that the field of force ($^{\Sigma/T}F_I$) converges φ_+, whereas the field of energy ($^{\Sigma/T}E_I$) asymptotically approaches φ_+.

Close inspection reveals that although both fields have 24-fold analog polarity transformations, this 24-fold pattern is only a third of a larger analog cycle in the sense that 72 intervals are required for a neutral inversion center to align with a neutral interval value (I=0 and I=72 are neutral intervals that coincide with neutral inversion centers of the 24-fold sequence). It may be coincidental, but half of this longer cycle (from intervals 0 to 35) corresponds with the convergence of the Lucas series on a near ideal value of the golden ratio, and this value corresponds with the ratio of the 36th to 35th terms of the Fibonacci series. The relationship points to a convergence of the two fields on $\boldsymbol{\varphi}_+$ at interval number 35:

AS **I** INCREASES $^{\Sigma/T}F_I^{-I}$ CONVERGES ON $\boldsymbol{\varphi}_+$, THEREFORE…

$$\varphi_+ \approx {}^{\Sigma/T}F_{35}^{-35} = 20633239^{-35} \approx \frac{^{\Sigma/T}E_{35}}{^{\Sigma/T}E_{34}} = \frac{9227465}{5702887}$$

8.2.3 INTERLEAVED COMPLEMENTARY CUBOCTAHEDRAL DIPOLE FIELDS

All forms of continuous proportion are preserved through the continued maintenance of a fixed relationship between three consecutive terms. Drawing on the symmetry of the triangular dimension for a geometric analogy, consider the three triangular plane orientations that form cuboctahedral spherical close packing. In this analogy The Receptive Field is associated with planes A (past), B (present), and C (future). The Creative Field then becomes analogous to an interleaving of two sets of the same three planes (A_1+C_1), (B_1+A_2), (C_1+B_2), then repeating (A_2+C_2), etc. thereby suggesting two alternate means of constructing the same spherical close-packed symmetry. In the later case, the paired past and future close-packed planes coordinate the symmetry of the present plane thereby dictating its close-pack symmetry. By extending this analogy to the case at hand it becomes clear that symmetry transformations of energy are associated with a continuous proportion involving three terms whereas the symmetry transformations of force are associated with a continuous proportion involving six paired terms. In this way the force proportion coordinates and thereby dictates the symmetry of the energy proportion.

This interleaving symmetry can be extended indefinitely. As shown in Figure 35, when one repeatedly sums past and future polarities twelve tandem cuboctahedral dipole fields are evident, beyond which the polarity pattern repeats. The neutral inversion centers of every other field occur at the same interval. This suggests that there are two pairs of six fields: one set of six that compose facets of the complementary cuboctahedron aligned with the *energy* perspective, and the other set of six aligned with the *force* perspective. Consistent with the prior proposal that the dipole field of *force* is the derivative of the dipole field of *energy*, the interleaving fields beginning with the field of force followed by all those to the right of it are defined as successive "derivative" fields. Interleaving dipole fields can also be extended in the direction of the "integral" symmetry of energy. The fields beginning with energy followed by all those to the left of it are defined as integral fields in Figure 35.

The first "integral" of the energy field on the far left exhibits the same polarity transformations as the field just before the right-most field of the figure (a 5-4 dipole field of force). Although unlike the whole number "derivative" fields to the right of the energy dipole, the "integral" polarity fields to the left necessarily contain fractions. This first fractional field is of particular interest to me because I'm interested in determining whether or not it is possible to characterize our matter-centric perspective by extrapolation from the relationship between the dimensional super-symmetries of energy and force.

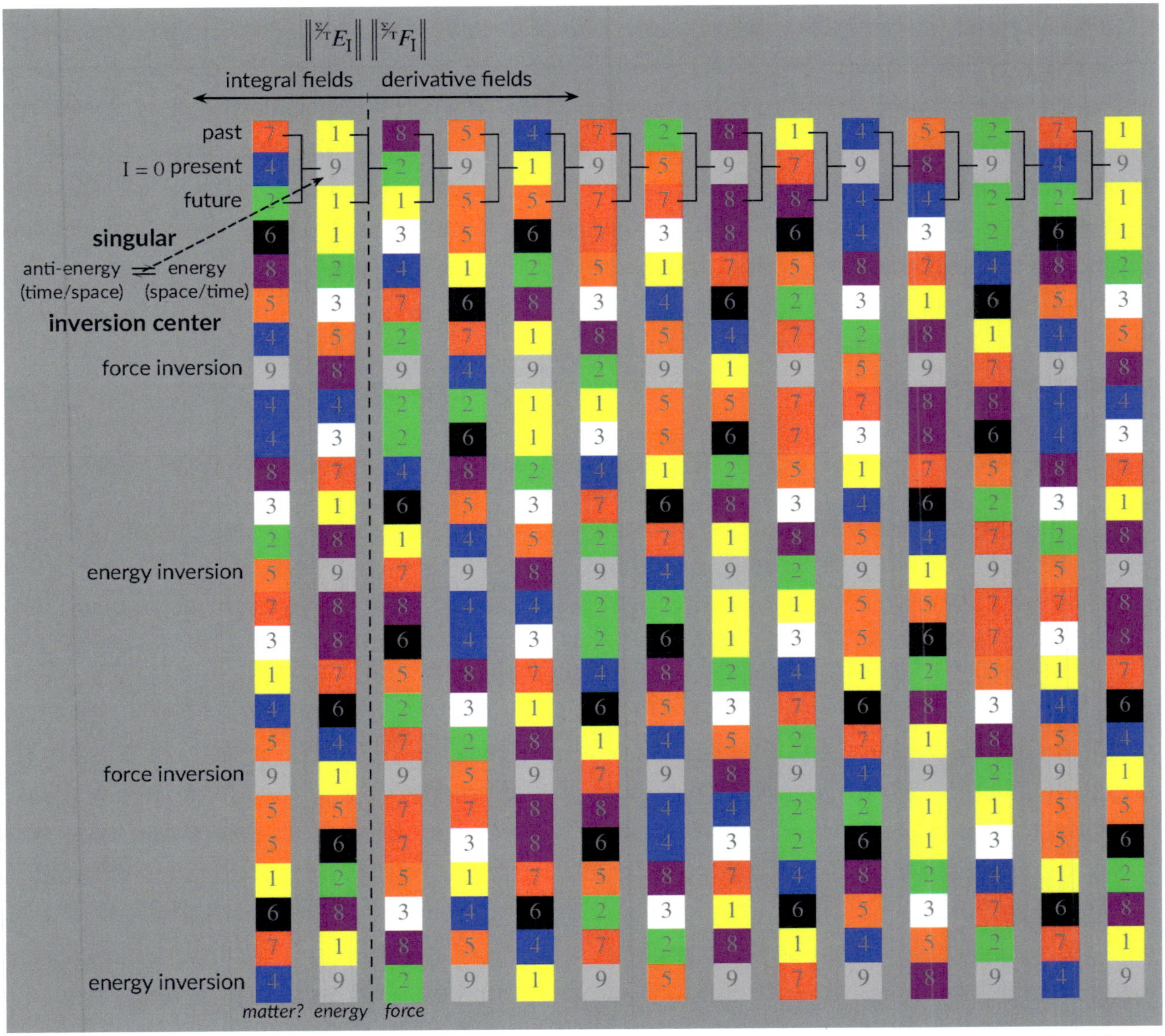

Figure 35. Interleaving cuboctahedral dipole fields of continuous golden proportion: "Derivative" fields are shown to begin with *force* and extend to the right. "Integral" fields are shown to begin with *energy* and extend to the left. This figure was generated by repeated summation of past and future intervals to form the present interval of the next dipole field to the right. This is shown graphically in the figure at interval zero where the past and future terms of the preceding proportion are summed to form the interval zero term of the succeeding proportion.

I will show in the next chapter why I believe the periodic energy state filling symmetry of atomic elements suggests that our matter-centric perspective is an integral perspective of energy. Given this, it is interesting to consider the possibility that this 5-4 dipole field of force might be connected to our matter-centric perspective since it is the first "integral" dipole field of energy according to this interpretation the entwined dipole field spectrum shown in Figure 35.

The idea of our matter-centric dimensional perspective being linked to the super-symmetry of the 5-4 dipole is an intriguing possibility, especially when I consider my own microcosmic symmetry. On the other hand, perhaps it's just a convenient extrapolation for me to think of these different cubic rotation axes within the super-symmetric cuboctahedral dimensional continuum as successive dimensional perspectives of the same underlying energy symmetry. This particular numeric analysis highlights interesting echoes of energy and force super-symmetry, but it is inconclusive.

8.3 CONVERGENT PHILOSOPHIES OF FIELD AND ANTI-FIELD

> The interpretation of *yinyang* as *qi* conceives *yinyang* as a dynamic and natural form of flowing energy, a complementary in the primordial potency of the universe. The *Huainanzi* offers more detailed explanation of the cosmological process of *yin* and *yang*:
>
> > When heaven and earth were formed, they divided into *yin* and *yang*. *Yang* is generated [*sheng*] from *yin* and *yin* is generated from *yang*. *Yin* and *yang* mutually alternate which makes four fields [*wei*, "celestial circles"] penetrate.[60]

As mentioned previously, the eastern philosophic school of thought that establishes *yin* and *yang* as complementary primordial symmetry elements of energy (*qi*) meshes will with the prior numeric analysis, which suggests that the golden roots of space and time (the base and anti-base of the continuous golden proportion) govern the super-symmetric multi-dimensional expansion of the energy field. The notion that *yang is generated from yin and yin is generated from yang* coincides with how spatial and temporal dimensions differentiate since continuous spatial proportions are incrementally propagated by sequential temporal constraints and temporal continuous proportions are incrementally propagated by sequential spatial constraints. Further still, the notion that *yin and yang mutually alternate which makes four fields* has obvious parallels with the cuboctahedral dipole field symmetry. Other texts go further to suggest that *there are six heavenly influences* on vital energy (*qi*) that produce such things as colors.[61] By highlighting these parallels I don't mean to suggest that ancient Chinese philosophers made the association between *yang* and space or *yin* and time nor do I suggest that they had any first hand knowledge of spatial-temporal cuboctahedral dipole inversion fields. However, I find it difficult to dismiss these parallels as purely coincidental. It seems more likely that there are many paths that lead to the same one-truth of why everything is as it is.

In ancient Chinese wisdom the two primordial aspects of nature always act in tandem resulting in dynamic inversion between their complementing qualities. This is suggestive of the symmetry of continuous golden proportions, which are described mathematically as being the combination of equivalent powers of spatial base and temporal anti-base. For the most part, then, the philosophic concept of *yin-yang* as vital energy doesn't separate thesis (*yang*) and anti-thesis (*yin*) into their unblended fractional primordial natures, yet it does appear to recognize that they exist as blended components that form a dynamic whole inverting system.

As primordial un-merged fractional forms, what are thesis and anti-thesis? That is, what are the pure symmetry elements that participate in the primordial *first cause* of why everything is as it is: pure-*yang* and pure-*yin*. I've suggested that pure-*yin* is analogous to the temporal anti-base and pure-*yang* the spatial base of continuous golden proportions, with the former being introverted and the latter being extroverted singular dimensional torus-like wisps of time and space. But space and time cannot exist independently; rather, they combine to form the spatial/temporal field of energy and the temporal/spatial field of anti-energy. Figure 36 is intended to aid the reader in considering the dynamic interplay between the primordial spatial and temporal symmetry elements that govern the continuous golden proportions of force and energy. Spatial and spatial/temporal field symmetries are juxtaposed with the temporal and temporal/spatial anti-field symmetries.

Notice that the dipole field of force (center) positively merges the pure subjective and pure objective symmetry elements of time (*yin*) and space (*yang*). The continuous force proportion represents the combined influence of spatial and temporal symmetry elements and the combined influence of past and future energy states on the present energy state. Whereas the continuous energy proportion contrasts future and past since the present interval is always there difference: the perspective of anti-energy (left) arises from the symmetric influence of a difference-relationship (separation) between time and space (time minus space), and the perspective of energy (right) arises from the symmetric influence of a difference-relationship (separation) between space and time (space minus time).

It is also possible to define an anti-force perspective, but only if one assumes the subjective perspective of the origin, effectively inverting objective and subjective symmetry elements. This would discount any inherent distinction between the two symmetry elements. So the perspective of force, as the positive combination of base and anti-base, seems naturally aligned to the spatial/temporal perspective, which may be the explanation for the natural orientation of space over time as dictated by the sequential ordering of spatial and temporal constraints.

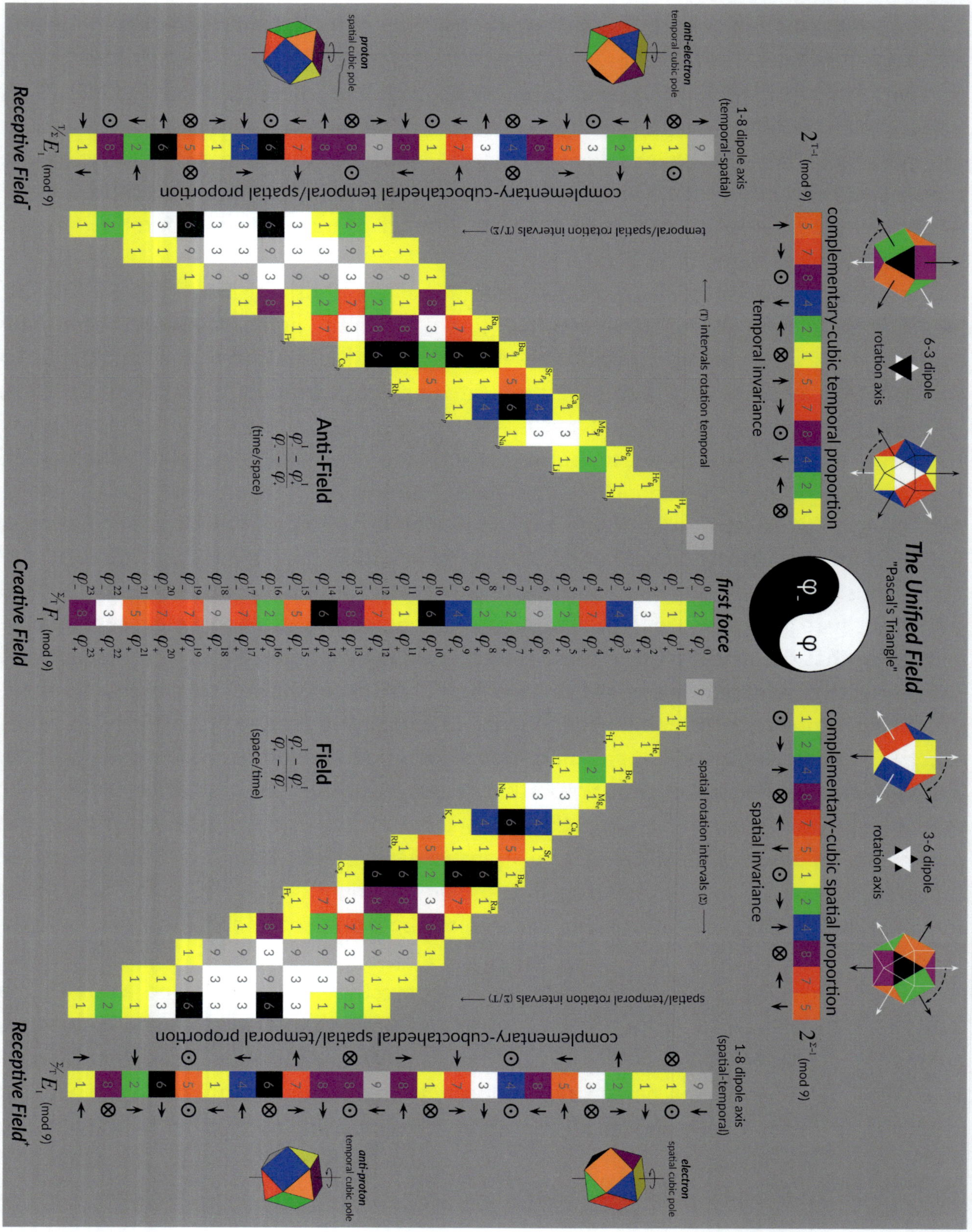
The Unified Field
"Pascal's Triangle"
first force
6-3 dipole
rotation axis
3-6 dipole
rotation axis
complementary-cubic temporal proportion
temporal invariance
complementary-cubic spatial proportion
spatial invariance
2^T-1 (mod 9)
2^Σ-1 (mod 9)
1-8 dipole axis (temporal-spatial)
1-8 dipole axis (spatial-temporal)
complementary-cuboctahedral temporal/spatial proportion
complementary-cuboctahedral spatial/temporal proportion
temporal/spatial rotation intervals (T/Σ)
(T) intervals rotation temporal
spatial rotation intervals (Σ)
spatial/temporal rotation intervals (Σ/T)
Anti-Field
(time/space)
Field
(space/time)
anti-electron
temporal cubic pole
proton
spatial cubic pole
electron
spatial cubic pole
anti-proton
temporal cubic pole
Receptive Field−
Creative Field
Receptive Field+

Figure 36. Field and anti-field symmetry: Spatial and spatial/temporal rotational perspectives of the origin of the energy field are shown juxtaposed with mirror temporal and temporal/spatial rotational perspectives of the origin of the anti-energy field. The dipole field of force (center) is the continuous golden proportion in harmony with sums of equivalent powers of the subjective temporal anti-base (*yin*) and the objective spatial base (*yang*). In contrast the fields of energy (right) and anti-energy (left) are the continuous golden proportions in harmony with difference-relationships between equivalent powers of base and anti-base, as shown.

> ...above these two principles [limited and unlimited] they [the Pythagoreans] have posited a unifying cause, superior to everything; which according to Archenetus, is the cause before the cause, and, according to Philolaus, the universal principle.
>
> Archytas [62]

This leads me to consider the philosophic and scientific question of a *first cause* or first symmetry. Here again I'll leverage the notion that *a thing governing is senior to a thing governed* set forth in Plato's Epinomis. Recall that since energy governs matter, I said that energy is dimensionally senior to matter. Using this same reasoning force is senior to energy since energy is governed by force.

By straddling the neutral inversion center of the origin the first force state of the Creative Field contains the germinal symmetry elements necessary to initiate and perpetuate the ensuing dimensional symmetry of energy. This *first force* is the sum of the golden base (space) and the golden anti-base (time) both raised to the null power:

$$\text{The } \textit{first force} \text{ of void} \;=\; \varphi_+^{\;0} + \varphi_-^{\;0}$$

This first symmetry contains the minimum three terms necessary to propagate the continuous golden proportions of the dimensional continuum of energy including all subordinate symmetries contained therein. It contains the symmetries of point and void and thereby has the seminal capacity to be the origin of all the geometric and numeric symmetries of space and time that arise there from.

> The universe is one, and it began to come into being from the centre, and from the centre upwards at the same intervals (of distance) as those below. For the parts above from the centre are in inverse relationship to those below; for the centre is to what is below as it is to what is above, and so with all the rest; for both stand in the same relationship to the centre, except in so far as their positions are reversed.
>
> Philolaus [63]

Time and space cannot exist independently of one another; consequently the dual perspectives of anti-energy and energy cannot exist independently of one another either. Yet, given that our reality is naturally aligned to the spatial/temporal perspective of the field of energy, it's not well reasoned to expect the field of anti-energy to appear as it does from the subjective temporal/spatial perspective of the origin shown in Figure 36. This begs the question: How does the anti-field appear from the spatial/temporal perspective of the field?

As a means of entering into this line of inquiry, I'll begin with a passage data-mined from another ancient philosophical tradition that provides some relevant metaphorical references to dimensional number symmetry: the ancient Zoroastrian creation myth. Porphyry wrote that the greater part of Pythagoras' wisdom was acquired during his time traveling in the near East, which included time with the Chaldeans (neo-Babylonians) and may have even included time in Persia studying Zoroastrian wisdom.[64] So the Zoroastrian creation myth is particularly relevant to the discussion at hand because the myth is very likely derived, at least in part, from the same ancient number theory that influenced Pythagorean thought.

> ... Revelation is the explanation of both *spirits* together; one is he who is independent of unlimited time, because Aûharmazd ... ever will be; *while* Aharman in darkness, with backward understanding ... *will* not be; And between them was empty space... Both are limited and unlimited spirits, for the supreme is that which they call endless light, and the abyss that which is endlessly dark, so that between them is a void...
>
> Bundahis – Zoroastrian creation myth [65]

The creation myth of the Zoroastrians depicts *endlessly light* and *endlessly dark* (*backward*) opposing symmetry elements with a *void* separating them. The *light* element, *one*, is complemented by the *dark* antithesis. Both elements are described as having *limited* and *unlimited* (*peras* and *apeiron*) qualities consistent with the qualities of continuous proportions referenced in Greek philosophy. Filtering out the embellishments, the *endlessly light* (thesis) and the *endlessly dark* (anti-thesis) are seemingly analogous to the notion of a field and an anti-field separated by a mirroring void lying between them.

Further parallels can be drawn between the thesis – who *ever will be*, and anti-thesis – who ever *will not be*, and the dominant field (spatial/temporal) and subservient anti-field (temporal/spatial). The universal spatial/temporal perspective being dominant over the temporal/spatial perspective owing to the incremental nature of constraint imposition that aligns with the natural perspective of force. So, although both perspectives exist, it appears that nature has chosen space over time. Therefore, when the anti-field is viewed from the perspective of the field it doesn't appear as it does when viewed from the perspective of the origin of the anti-field. Rather, from the perspective of the field, the anti-field appears *backward*.

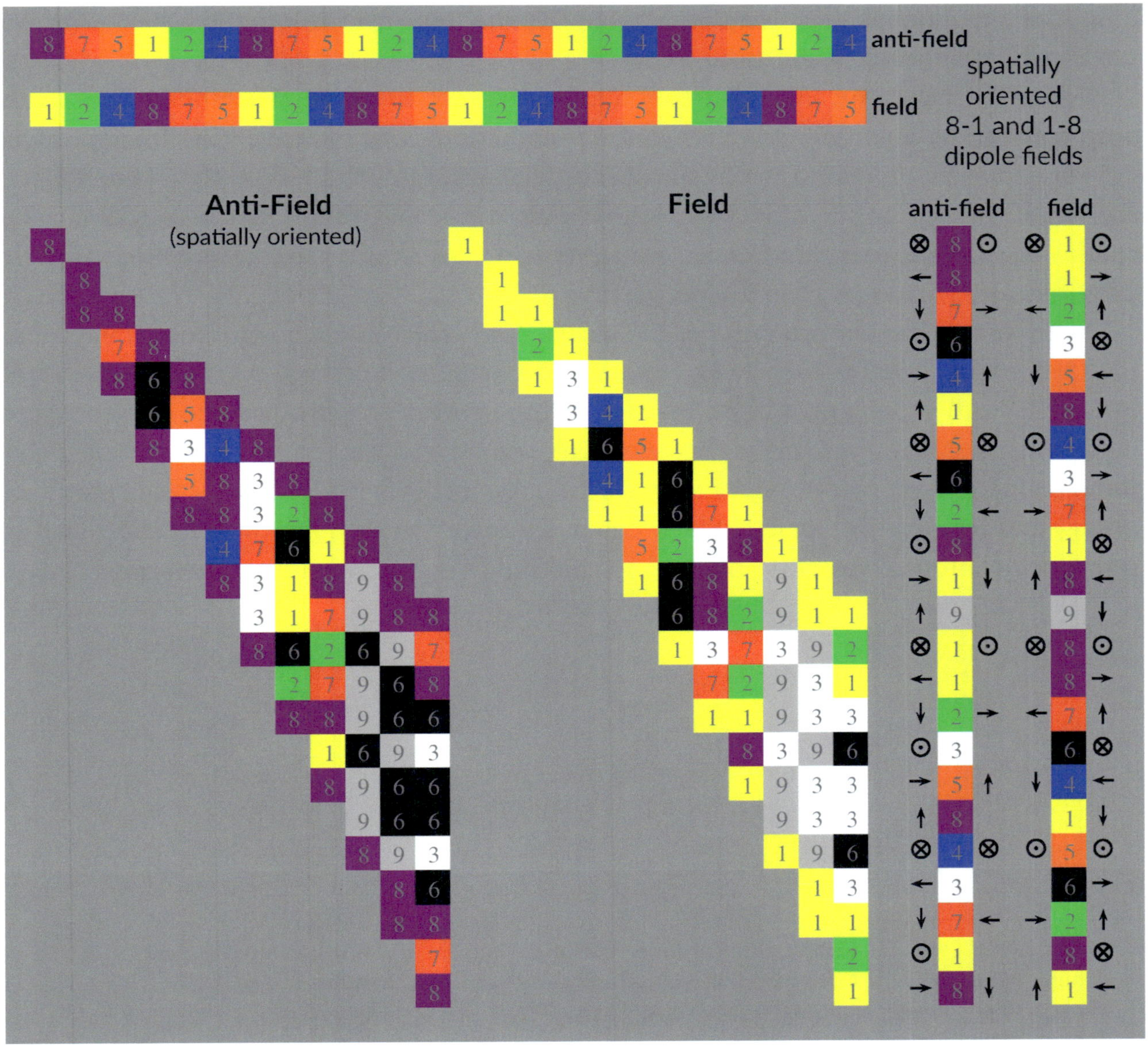

Figure 37. The anti-field from the perspective of the field: the individual translational symmetry states are converted to complements, the CCW temporal complementary-cubic proportion becomes CW rotation of complements; and the complex complementary-cuboctahedral temporal/spatial rotational proportion orients to its spatial cubic pole (8).

As the temporal/spatial perspective is the negative reciprocal of the spatial/temporal perspective and as the former is observed from the latter, the temporal/spatial perspective has to be converted to modulus-9 complements in order to show how it appears from the spatial/temporal perspective. Notice that temporal/spatial rotational perspective of anti-energy is a co-constrained temporal-spatial 1-8 cuboctahedral dipole field, which means that the cubic polarity-8 is spatially polarized. I propose then that from a spatial/temporal perspective the temporal/spatial anti-field is oriented such that we observe its spatial pole (polarity-8).

Figure 37 shows the anti-field reoriented to the spatial/temporal perspective, which is accomplished by inverting polarities to their modulus-9 complements. First the individual field states associated with translational perspectives are converted to complements and reoriented to conform with spatial/temporal constraint ordering, then spatial and spatial/temporal constraint intervals are summed to reveal the super-symmetric anti-field polarities as they appear from the perspective of the field.

Note that if one were to observe the virtual complementary dipole super-symmetry of the anti-energy and energy fields together from a spatial/temporal perspective that integrates both incrementally, just like atomic matter integrates protons and electrons into its quantized states, they might very well look like finite polarized spinning *particles*. This thought leads next to the validation of the proposed dimensional super-symmetry of energy and anti-energy dipole fields by comparing their symmetry to the periodic quantum symmetry of atomic elements and their constituent energetic point-charge *particles*.

CHAPTER CITATIONS

[58] Huang, Alfred 1998 *The Complete I Ching* Inner Traditions International (Vermont) p 41
[59] Huang, Alfred 1998 *The Complete I Ching* Inner Traditions International (Vermont) pp 23-24
[60] Wang, R. http://www.iep.utm.edu/yinyang/#H3
[61] Wang, R. http://www.iep.utm.edu/yinyang/#H3
[62] Guthrie, K.S. 1988 *The Pythagorean Sourcebook and Library* Phanes Press (USA) p 179
[63] Freeman, Kathleen 1948 *Ancilla to the Pre-Socratic Philosophers* Harvard Univ. Press (USA) p 76
[64] Guthrie, K.S. 1988 *The Pythagorean Sourcebook and Library* Phanes Press (USA) p 20
[65] trans. E. W. West 1880 *Pahlavi texts, Part 1* (*Sacred Books of the East* vol. 5) Oxford University Press pp 4-5

The Birth of Complexity

> … at present one should keep in mind three kinds: that which comes to be, that *in which* it comes to be, and that *from which* what comes to be sprouts as something copied. And what's more, it's fitting to liken the receiver to a *mother* [energy], the "from which" to a *father* [force], and the nature between these two things as an *offspring* [matter]…
>
> Plato [66]

As the field that is the *offspring* of *mother*-energy impelled by *father*-force, matter is junior to that which governs it. Being a copy, the offspring is less perfect than its progenitors. Yet as a result of these imperfections the offspring acquires complexity beyond that *in which* and *from which* it was conceived.

Atomic matter integrates spatially oriented anti-energy field dipoles (protons), energy field dipoles (electrons), and their neutral combination (neutrons) in numeric progression to satisfy the dimensional symmetry states of energy (*quantum states*). Elemental atomic energy states emulate the incremental symmetries of the singular point that is summed to form the translational dimensional proportions of Pascal's Triangle. Atomic matter is able to incrementally integrate these field and anti-field dipoles owing to the constraining influence of the finite regions of void that separate them.

> The followers of Pythagoras hold that there is a void outside the universe into which the universe breathes forth, and from which it breathes in.[67]

This integration of void between field and anti-field dipoles creates atomic matter from energy thereby consuming and filling the infinite emptiness of the vacuum. Each atom-localized quantum state is subject to the *Pauli exclusion principle*, which dimensionally excludes more than one dipole (*Fermion*) from occupying the same state concurrently. For instance no two electrons can occupy the same dimensional symmetry state in the same atom simultaneously.

Additionally, these dipoles cannot fully satisfy more than one atom-localized translational point-symmetry at a time. Within the atom the dipole super symmetry of the point must conform to the translational point-symmetry of the singular state it satisfies. Field states within the atom are split into their spatially and temporally oriented component symmetries. In contrast numeric species in the translational states of Pascal's Triangle are free to conform to differing spatial and temporal symmetric perspectives. For instance, the spatial triangular number fifteen is also the temporal pentangular number fifteen.

Dipoles cannot satisfy both spatial and temporal translational point symmetries simultaneously owning to the asymmetric dimensional constraints that define their natural orientation. As an apparent effect, each electron-proton pair must acquire neutral invariance (neutrons) to disallow symmetry conflicts between alternate spatial and temporal translational orientations. This symmetric requirement for atoms to integrate neutral invariance seems to be compounded by the freedom of the electron to be energetically promoted to higher states within the dimensional continuum. This freedom apparently influences the integration of multiple neutrons for a given electron/proton pair, causing neutron buildup and further symmetric instabilities. Yet these same symmetry conflicts are what enable the *offspring*-matter to express levels of complexity far beyond *mother*-energy and *father*-force.

9.1 Particle Theory Correlations

The dimensional symmetries of continuous proportions discussed in the preceding chapters provide qualitative explanations for such things as the sub-atomic *particle* symmetries observed when protons are broken apart by high-energy particle accelerator collisions. By design these experiments are intended to reveal the building blocks of matter. Yet the method is somewhat like destroying a stable structure so that you can get a fleeting glimpse at its underlying internal elements before it collapses. So, although the method provides an inkling of the symmetries involved in the underlying structure, the logic of structure itself remains elusive.

My intent in this section is to highlight where I believe these fleeting glimpses of symmetry originate, rather than to expound every flavor of symmetry observed through the process of disrupting stable dipoles. More importantly, I intend to describe the origin of the symmetry of the stable sub-atomic particles and to discuss the forces that govern their behavior in terms of dimensional symmetry. By no means will this discussion be entirely thorough and cover every flavor of particle in every branch of quantum theory, as this would require the efforts of someone more vested in that branch of science than I.

Rather, my intent is simply to expound upon the body of scientific observations that corroborate the natural complementary numeric interpretation of dimensional symmetry that has been detailed in the preceding chapters, since I believe it provides a more comprehensive and rational explanation for said experimental observations.

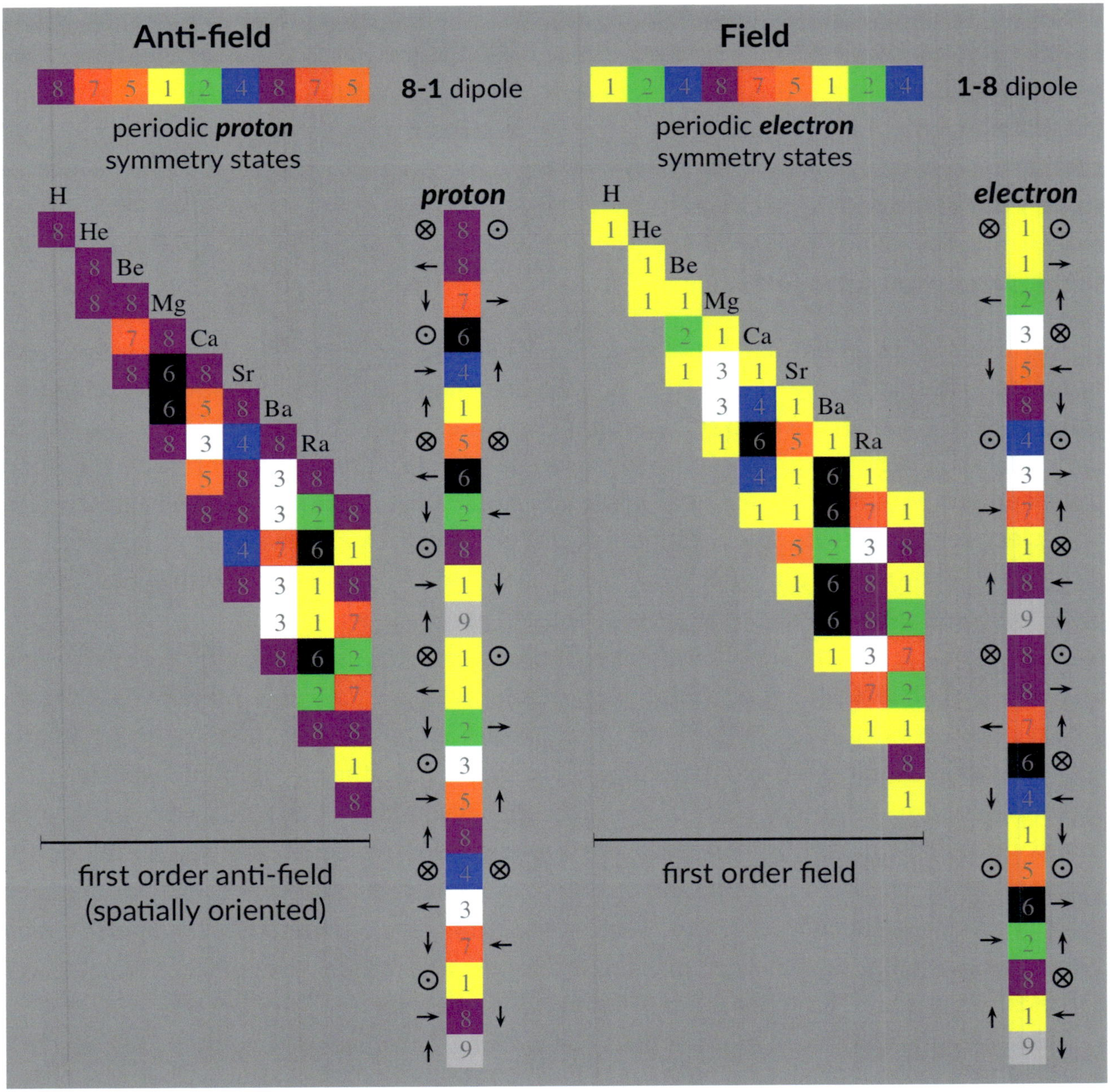

Figure 38. Spatial/temporal oriented anti-energy and energy fields and associated rotational super-symmetries: full isotropic s-shell elements are shown as the culmination of spatial intervals; the first order symmetry field is shown to correspond to the theoretical limit of the periodic table; electrons and protons are shown corresponding to complementary-cuboctahedral dipole symmetries.

I'll begin with Figure 38, which is essentially the same as Figure 37 excluding the labeling: the spatial/temporal orientation of the anti-energy field is associated with proton symmetry states, and the states of the energy field are associated with electron symmetry states. Notice also that spatial rotational intervals are associated with full spherical s-shell elements, which mark the completion of the integration of dipoles into all field and anti-field states within a given spatial interval (*period*). The first-order field, as observed from a spatial rotational perspective, corresponds with the theoretical 120 element bounds of the periodic table. However, the final first-order column lacks a full s-shell element label since the s-shell elements (119 and 120) that finish-out the first order field have not yet been synthetically manufactured. The details of how I've come to these conclusions will be spelled out in the coming sections.

9.1.1 FIELD AND ANTI-FIELD DIPOLES

From our matter-centric perspective all anti-energy and energy states are virtual until occupied. The electron, for example, is able to occupy any one of the allowed energy states within a symmetric hierarchy, but these states are not experimentally observable until they are filled. It follows that the electron has dipole super-symmetry, since it harbors the symmetric capacity to satisfy all dimensional symmetry states within the continuum. The electron appears to be a finite integral of energy super-symmetry that has the capacity to conform to the symmetry of any singular component of the magnitudes that form the continuous translational proportions of Pascal's Triangle.

This leads me to conclude that the *electron* expresses the symmetry of the spatial pole of the 1-8 spatial-temporal cuboctahedral dipole field (spatial cubic polarity-1), that the *proton* expresses the symmetry of the spatial pole of the 1-8 temporal-spatial cuboctahedral dipole field (spatial cubic polarity-8), that the *anti-proton* expresses the symmetry of the temporal pole of the 1-8 spatial-temporal cuboctahedral dipole field (temporal cubic polarity-8), and that the *anti-electron* expresses the symmetry of the temporal pole of the 1-8 temporal-spatial cuboctahedral dipole field (temporal cubic polarity-1). At minimum the proposed complementary cuboctahedral dipole super-symmetry has the capacity to explain the origins of these experimentally observed *particles* and it is hard to imagine a more rational explanation for their existence.

If we assume that the experimentally determined left-handed chirality of the electron arises from its cuboctahedral rotational super-symmetry, we can define the electron's axis to be the co-constrained rotation axis of the 1-8 spatial-temporal dipole. By correlating the electron's rotation axis with the dimensional symmetry of energy in this manner it is then possible to determine whether or not the arbitrary designation of spatial translational perspectives as occupying the columns and temporal the rows of Pascal's Triangle in Chapter 3 was correct.

By *right-hand rule* convention the left-handed electron should rotate on it's own axis in the direction of the curl of the fingers on your left-hand (clockwise). Per Figure 28, the spatial/temporal cuboctahedral polarity transformations begin in the counter-clockwise direction. But from an external fixed reference point, rotation of the cuboctahedron itself must be in the clockwise direction to observe counterclockwise polarity transformations. So it's a matter of perspective whether on not the original selection of handedness back in Chapter 3 is correct or not. If I take the perspective of the central point, spatial and temporal designations should be reversed. But if I take the perspective of an external observer, they need not be reversed. Either way, this is not a problem because spatial and temporal directions can be easily reversed with no effect on the subsequent analysis. I'll simply use the experimentally confirmed handedness of the electron as the reference and apply it to my interpretation of Figure 28 and beyond.

The chirality of the other dipole species can be determined by examining their dipole orientations relative to the left-handed electron: The chirality of the anti-proton is **left-handed** from the spatial/temporal perspective and right-handed from the temporal/spatial perspective since it is the opposite pole of the same spatial-temporal 1-8 dipole field as the electron. The chirality of the proton is **right-handed** from a spatial/temporal perspective and left-handed from a temporal/spatial perspective. And, the chirality of the anti-electron is **right-handed** from a spatial/temporal perspective and left-handed from a temporal/spatial perspective. For consistency I will reference the spatial/temporal perspective when designating handedness (shown in **bold**).* The proton and anti-electron are opposite poles of the same temporal-spatial 1-8 dipole field. Note that the proton and anti-proton are not currently known to have a specific handedness that I am aware of. So this is a new finding that I'll apply to the *weak interaction* in the next section.

If electrons, protons, and their anti-particles are complementary-cuboctahedral field and anti-field dipoles, what are the other generations of matter such as the *muon*, which behaves like an electron but is *heavier* (a more energetic version)? The analog nature of cuboctahedral dipole fields suggests that equivalent symmetry states exist every 24 intervals. This symmetry then predicts higher energy, symmetrically equivalent, dipoles. However, correlations between specific energies quoted in the literature and specific field states are beyond the scope of this qualitative analysis. What is important is that the symmetry supports what has been experimentally observed.

---

*The hands-on physical models provided in Appendix B are ideal for visualizing these rotational transformations.

9.1.2 NEUTRAL PARTICLES

Since it is true that in general all forms of neutrality arise from the constraint of void and that the void always separates the proton and electron unless both are summed to form a *neutron*, as 8+1=9, then it follows that neutrons are manifestations of invariance that arise from the constraint of void that separates proton-electron pairs. This neutral invariance appears to be associated with the neutral symmetry between a given pair of field and anti-field point symmetries. I've suggested that, given the restrictions on the matter perspective embodied by the *Pauli exclusion principle*, neutrons are required to constrain proton-electron pairs to a given translational point symmetry within the first-order symmetry field of Pascal's Triangle.

Since the electron exists as a cloud of negative charge that surrounds the proton with empty space seemingly captured between them, it seems reasonable to posit that this neutral invariance is associated with the sequestered void or emptiness that separates a given proton-electron pair.* Yet in the case of Hydrogen the symmetry states of the electron and the proton apparently don't require invariance integration (neutrons) since there are no symmetry conflicts possible with only one proton-electron pair. The symmetry conflicts beyond Hydrogen necessitate the acquisition of neutrons to act as sentinels that fix symmetric perspectives thereby disallowing symmetry conflicts. I consider the function of the neutron to be like a surrogate; similar to the way a person would engage a surrogate to hold a place in line.

The devolution of the neutron into its constituent parts includes an *antineutrino* as well as an electron and proton. This transformation is referred to as *beta decay*. Beta decay is inevitable for neutrons outside the confines of the atom. The average lifetime of a neutron outside the nucleus of an atom is roughly 15 minutes,[68] which suggests that its stability is associated with the symmetry state it occupies within the atom. I'll discuss these symmetry states in the section devoted to quantum numbers, so for this discussion it will suffice to say that neutron symmetry states can have spatial or temporal *spin* (±½) just like a proton state and similar to an electron state. Beta decay can also occur inside an unstable nucleus in situations where there are excess neutrons, in which case such atoms transform to the next higher atomic element. However, in general neutrons are considered highly stable inside stable nuclei. So I'll focus on the case of beta decay outside the nucleus.

---

*This interpretation meshes with the Pythagorean notion that there is a void that the universe *breathes in* since, by combining field and anti-field dipoles, matter consumes empty space. Therefore *the universe breathes forth* (expands) into the *void outside the universe* as matter is formed from energy.

The atom-localized symmetric force of neutral void that constrains the electron and proton to the coupled neutral neutron state is conventionally referred to as the *weak force*. The weak force is conceived of as an intermediate particle that is the combination of the electron and the antineutrino, namely the *intermediate vector boson* (W). This theoretical force particle is used to describe a two-step decomposition of a given neutron (*n*) into a proton (*p*), electron (e^-), and an anti-neutrino ($\bar{v}$) as follows[69]:

$$n \rightarrow p + e^- + \bar{v} \quad \text{becomes} \quad n \rightarrow p + W^- \quad \text{followed by} \quad W^- \rightarrow e^- + \bar{v}$$

In the preceding reactions both charge and angular momentum (spin) must be *conserved*, which means there can be no net gain or loss. In step 1 of the two-step decomposition the neutron transforms into a positive proton and a negative boson, and the charge is conserved since positive and negative charges cancel to neutral. But, given the new learning about proton chirality, what about spin conservation?

By convention the matter particles (*Fermions*) that form atoms such as electrons, protons, and neutrons occupy states with spins of ±½. Of these particles the electron alone belongs to a particle class called *Leptons*, which also includes the neutrino and two other generations of electron-like and neutrino-like particles and their anti-particles. Particles of this class are known to have left-hand chirality (*spin ½*), and their anti-particles are known to have right-hand chirality (*spin -½*). In contrast the W^- Boson, which is believed to mediate the above reaction, is considered a spin 1 force particle. And right-handed proton chirality (*spin -½*) is apparently not factored in.

The total angular momentum of a given particle is the sum of its spin on its own axis (its chirality) and the *spin* of the state it occupies within the atom (its *orbit*). These are different forms of spin, even though both are characterized as half integer spins. In fact the momentum associated with the atomic state the neutron *occupied* is already lost or in the process of being lost since it isn't in that state when it decays outside the atom. I don't think this observation is fully factored into the preceding equation otherwise the neutron wouldn't readily decay in the first place. In other words both quantum state momentum (orbital) and inherent dipole momentum (chiral) are factors to be considered. So I think it's unreasonable to assume that *all* the force holding the neutron together is suddenly released when it decays. I think part of the force is already gone. The part that is released in this reaction appears to me to be associated with the relative dipole rotational orientation of the proton and electron (chiral).

I take the integer spin associated with force to mean that it is the sum of energy (electron) and anti-energy (proton) rotational orientations. Since the electron's spin is left-handed and the proton's spin is a right-handed anti-spin, by convention their spins sum to

1 just like the electron and antineutrino in the preceding reaction.* Implied then is that at the time of decay the spin of the neutron is 1 and that in a stable state it is 1-chiral ±½-orbital.

According to my interpretation of this dipole symmetry interaction the *intermediate vector boson* is a means of describing the chiral component of the neutron's momentum, since the W^- boson is composed of a left-handed electron and a right-handed antineutrino whose chirality matches that of the proton (step 2). Therefore, I interpret the antineutrino to be a residual integral waveform (a wave-*packet*, *bundle*, or *particle*) generated as the symmetrical effect of the loss of neutrality when a neutron decomposes into a proton and an electron. And I think antineutrino emulates the chirality of the proton versus the electron because the electron is bound to the proton, not vice versa.

By this interpretation the antineutrino is the residual constraint that binds the electron to the proton outside the atom. But the neutron looses its dimensional context outside the atom, which suggests that the dimensional invariance (constraint) between field and anti-field states within a given atom is responsible for the neutron's stability. In conventional terms the binding energy of the neutron to the nucleus is greater than the energy released from the decay of neutron, which accounts for it's stability when bound to the nucleus. Hence, I'm suggesting that the source of the binding energy is the dimensional constraint of complementary field states held invariant by the neutron.

9.1.3 Dismembered Dipoles

The dimensional number theory presented thus far provides a geometric first principles explanation for the existence of the proton. Therefore it isn't necessary to inquire further as to how a proton is held together since we have observed the spatial and temporal dimensional constraints that form its geometric structure.

Modern particle physics attempts to characterize the structure of the proton by analyzing the energy and symmetry of its perceived constituents – artifacts of particle accelerator collisions. The American physicist Murray Gell-Mann developed a theory for organizing the ever-growing myriad of observed products of these collisions (*mesons* and *baryons*) into octets, which he called the Eightfold Way. Wherein he used linear algebra to organize the products of these collisions into what are known as Gell-Mann matrices, which revealed that they could all be described by a common SU(3) symmetry. This led to the development of the quark model and later quantum chromo-dynamics (QCD).

---

*This meshes well with how I defined the perspective of force previously (the combined influence of objective and subjective symmetry elements).

These models suggests that there is a *strong force* that holds a triad of elementary particles together called *quarks* that form the proton. These *quarks* are said to be sometimes *down* or *up*, and at other times *strange* or *charm*ing, and sometimes on the *bottom* or *top*, though *strange* and *bottom* ones are considered *down*-like and *charm*ing and *top* ones are considered *up*-like. Down-like quarks are said to have electric charges of -1/3rd and up-like +2/3^{rds}. Hence, when an electron (charge -1) is added to an up-quark it changes it to down-like (2/3-1=-1/3) and in so doing a proton containing two ups and a down (udu) is converted to a neutron composed of two downs and an up (udd). Further, these quarks are said to carry color charges likened to the red-green-blue (RGB) of many consumer electronics devices.

These *color charges* correspond in general with the complementary chromo-numeric dimensional super-symmetries of the proton, but they are fleeting de-composite symmetries. I will not attempt to explain the sources of each unique flavor of quark, as it is sufficient in my mind to point to the obvious complementary dimensional symmetries that have the capacity to explain these experimentally observed component sub-symmetries – namely the mirroring SU(2)-like spatial temporal symmetry, the complementary cubic SU(6)-like symmetry, and the complementary cuboctahedral SU(24)-like symmetry described herein.

I think that the science of analyzing particle collisions to understand the building blocks of nature is at or near the detection limits of the technology. Only so much can be gleaned from bombarding the structure of field and anti-field dipoles since the symmetry of the dipoles is either disrupted or destroyed in the process. Although it is certainly understandable why this method has been employed. I simply think it has reached the limits of its usefulness as an analytic technique.

In my view these questions can be more readily answered by seeking to understand the first principles of dimensional symmetry, which suggests that the strong force that sustains the symmetry of the proton is the symmetric force of succeeding dimensional constraints that create its complementary cubic and cuboctahedral dipole fields. This force of symmetry can be traced back to the orthogonal juxtaposition of the germinal symmetry elements of point (1) and void (0).

9.1.4 THE UNSEEN HAND OF SYMMETRY

In the case of the force that enables an electron to acquire a higher energy state, as with an electron moving above its valence state within the confines of an atom, the motivating force is conventionally referred to as a *photon*, a discrete quantity of *electromagnetic force* which is often likened to the momentum of a charge-less, mass-less, particle traveling at the speed of light (a *force particle*). When an excited electron returns to its resting valence state it emits a photon with a discrete frequency (1/wavelength). This implies that atomic energy state transitions are influenced by some sort of symmetric size constraint on the transitions between energy states since these atomic energy state transitions involve electromagnetism. Further, how is it that this emitted photon can travel great distances without frequency deviation* and be absorbed by another electron in the equivalent valence state unless there is some sort of medium of symmetry to sustain its discrete momentum-fixing waveform? Plank's experimental observation of the constant relationship between the frequency of a photon and the energy its momentum imparts seem to confirm a size constraint on the field state transitions of dipoles.

So far I've focused on virtual energy symmetry of singular size and its resultant dipole fields that have the capacity produce *point charge particles*. An axiomatic theory of electromagnetism requires comprehension of another numeric symmetry: the symmetry that mediates changes in dipole field states. In order to understand the dimensional symmetry of electromagnetism I will explore how changes in time (frequency) and space (wavelength) could be triangulated by a specific form of continuous spatial-temporal prime number proportion that has the capacity to mediate changes in the energy states (*angular momentum*) of cuboctahedral dipoles. However, before I present the reader with the continuous dimensional proportions that I believe have the necessary symmetry elements to mediate these state changes, I want to finish examining the nature of atomic elements, beginning with the quantum numbers that characterize their periodic symmetry.

9.2 Quantum Number Correlations

> ... those called Pythagoreans applying themselves to the sciences, first developed them; and being brought up in them they thought that the first principles of these (i.e. numbers) were the first principles of all things.
>
> Aristotle [70]

9.2.1 Conventional Quantum Numbers

Conventionally four forms of quantum numbers are leveraged to define the periodic symmetry states occupied by electrons within atomic elements: principal, orbital, and magnetic and spin. The *principal quantum number* was developed by Bohr to characterize energy levels of the Hydrogen atom. His model is based on the assumption that electron

---

*With all else being equal, i.e. excluding source and detector velocity effects (red-shift/blue-shift) and other intervening influences such as gravitational fields or other optical effects like dispersion.

orbital energy levels in Hydrogen are discrete (*quantized*) owing to characteristics analogous with the wave-like behavior of harmonic oscillators*, where integer nodes of constructive interference are observed (resonant frequencies). Ultimately this led to periods of the periodic table (rows) being associated with integer radial distances from the nucleus – a proton in the case of Hydrogen. Notice that this is a one-dimensional model (radius).

These principal-radii inspired the notion of concentric *shells* and *sub-shells* of states associated with each period of atomic elements. Each period is distinguished by its principal-radius, i.e. by its principal quantum number. Yet Bohr's model didn't explain the presence of sub-shells. So Schrödinger followed Bohr's lead and explained these sub-shells by expanding the Bohr model into three-dimensions using spherical polar coordinates. He theorized that in addition to the radial component associated with the principal quantum number there were two angular components (latitude and longitude) that contributed to the oscillating waveform. Schrödinger then applied Newton's laws of conservation of energy and momentum to his hypothetical three-dimensional harmonic waveform: using partial differential equations, he found the possible three dimensional geometries, i.e. the possible radii as well as the possible latitudinal and longitudinal components that could sustain such a spherical harmonic waveform, all of which were integers (*quantized*).

The radial solutions associated with shells or periods were the same as Bohr's: namely the principal quantum number (*n*), which may be any positive integer. The latitudinal component was associated with sub-shells or orbitals: the orbital or azimuthal quantum number (l), which may be any integer from zero to *n*-1. Yet for the periodic table this quantum number is limited to 0, 1, 2, and 3 corresponding to the spectroscopic† notation s, p, d, and f. The longitudinal component was associated with the states an electron may occupy within a given orbital: the magnetic quantum number, whose integers range from – l to + l conveniently matching the number of electrons of each spin that each sub-shell may contain.

The spin quantum number was devised afterwards to explain spectroscopic observations of electron energy state splitting. Herzberg observed that the existing quantum numbers were *no longer adequate for He* [helium] *and the alkaline earths, since for these there are twice as many term series as for alkalis.*‡ [72] In other words, where the existing theory had predicted 1 energy state he found two such states separated by a small energy difference. Furthermore Zeeman observed that these energy differences could be

*String instruments such as Pythagoras's *monochord* used to develop his theory of harmonic proportions.[71]

†Spectroscopy in this context is the science of measuring the wavelengths of photons emitted by atoms and equating those wavelengths with electron energy state transitions.

‡Alkaline earths have two s-orbital valence electrons and alkalis have one s-orbital valence electron.

changed under the influence of a magnetic field (the *Zeeman Effect*). Ultimately the splitting was attributed to electrons in each orbital being paired *spin-up* and *spin-down*.

Free electrons were observed always to have the same spin or inherent momentum (chirality), which was assigned the quantum number ½. Yet it was theorized that these free electrons could orbit the nucleus in either the same direction as their spin or in the opposite direction. When an electron is envisioned hypothetically orbiting the nucleus in the same direction as its inherent spin, the axis of its own spin and that of its orbit are considered parallel. Such an electron is considered to be in a spin-up state (½). When an electron is envisioned hypothetically orbiting the nucleus in the opposite direction, the axis of its own spin and that of its orbit are considered anti-parallel, and such an electron is considered to be in a spin-down state (-½). These two electrons, then, don't violate the Pauli exclusion principle because they differ in the quantum states of their spins.

9.2.2 The Golden Waveform: The Dimensional Number Interpretation

As is evident, I have an alternate interpretation of the data. I don't agree with the assumption that the waveform of the electron stems from harmonic oscillation. I think dimensional symmetry clearly indicates that this is a false interpretation of the data. The harmonic oscillator model mimics reality since it was incrementally developed to explain experimental data, but it is not supported by axiomatic first principles. Whereas the dimensional number theory I've proposed not only correlates precisely with observation, it also provides rational geometric first principles that explains those observations. The conventional spherical harmonic oscillator model and the dimensional number symmetry model that I propose do agree to the extent that they both predict spherical waveforms for s-orbital electron states, since dimensional number theory predicts singular spherical symmetry for all such states.

The harmonic oscillator model, however, has led to many false assumptions. For instance, conventional quantum numbers have led to a dismissive explanation for why "4s" sub-shells are filled prior to 3d. But, as I will show, this paradox can be resolved by reinterpreting quantum states as spatial-temporal symmetry states governed by the axiomatic first principles of Pascal's Triangle. So in this section I'll compare and contrast quantum numbers derived form the conventional harmonic oscillator model with those I've derived from spatial-temporal dimensional number symmetry. Dimensional number theory provides qualitative explanations rather than the quantitative approximations that conventional quantum mechanics provides, like for instance the difference between electron energy levels in Hydrogen's sub-shells. This form of inquiry is not something that is within the scope of the present discussion since it requires an understanding of the dimensional symmetry that has the capacity to mediate field state transitions (next chapter).

SPATIAL INTERVAL QUANTUM NUMBERS: *Principal Quantum Numbers*

I've indicated that principal quantum numbers, which define the *periods* or rows of the periodic table, equate with the super-symmetric spatial intervals of Pascal's Triangle. However, due to the order in which states are filled, the singular spherical states (s-orbitals) are the last to fill and therefore belong on the far right of the periodic table, not the far left as is common with most conventional tables (see figure 41).* Alkaline earths are shown above the spatial intervals in Figure 38 to indicate that they are the final or culminating elements of a given spatial period.† The differentiated symmetry states of a given spatial interval appear to be incrementally satisfied by the integration of electrons, protons, and neutrons thereby neutralizing symmetry gradients culminating in the periodic return to the isotropic singular *s-orbitals*, which serve a the basis upon which the next period of symmetrical differentiation occurs.

The result of placing s-orbitals at the end of a given period rather than the beginning is that the period numbering of orbitals other than s-type increases by 1. For instance 2p is now 3p based on this interpretation. The present quantum numbering convention leads to the following filling order: 1s, 2s, 2p, 3s, 3p, 4s, **3d**, 4p, 5s, **4d**, 5p, 6s, **4f**, **5d**, 6p, 7s, **5f**, **6d**, 7p (*inner* out-of-order sub-shells are shown in **bold**). Numbering *shells* by spatial interval yields the following filling sequence: 1s, 2s, 3p, 3s, 4p, 4s, 5d, 5p, 5s, 6d, 6p, 6s, 7f, 7d, 7p, 7s, 8f, 8d, 8p, 8s‡. Notice that there are no longer any *inner* orbitals that require rationalizations such as *the 4s orbital becomes more stable and fills before the 3d.*[73] Notice also that the much-debated placement of Helium is resolved with the proposed *Spatial Interval Quantum Numbers*.

SPATIAL-TEMPORAL ORIENTATION QUANTUM NUMBER: *Spin Quantum Number*

The spin quantum number is used to distinguish between the paired spatial-temporal symmetry states that electrons satisfy for a given translational dimensional orientation (orbital type). As discussed previously, the electron is left-handed due to the rotational symmetry transformations of the 1-8 spatial-temporal cuboctahedral dipole. Yet the electron may occupy either spatial or temporal translational symmetry states: an electron

---

*The rules governing filling order are related to the prior discussion on incremental constraint imposition.
†Departing from convention, I've added a period 0. Hydrogen is present in both period 0 and period 1.
‡This sub-shell is predicted by symmetry, but these elements have not yet been synthesized.

in a *spin-up* state occupies a spatial translational symmetry state aligned with its natural spatial-cubic dipole orientation, and an electron in a spin-down state occupies a temporal translational symmetry state in opposition to its natural dipole orientation.

In conventional terms this is interpreted to mean that in a spin-up state the angular momentum of an electron (spin on its own axis) is aligned parallel with the orbital angular momentum of the state it occupies (its direction of orbit about the nucleus). Conversely, an electron in a spin-down state orbits the nucleus in the opposite direction, so its spin on its own axis is in the opposite direction of its orbit (anti-parallel). The notion of the electron orbiting the nucleus has provided many useful analogies. However it doesn't accurately depict what is going on physically. Merging previous observations of dimensional symmetry with conventional theory suggests a physical model for the electron akin to a symmetric waveform distributed about the nucleus that conforms to the spatial or temporal translational energy symmetry state it occupies. Therefore I consider spin quantum numbers to be *Spatial-Temporal Orientation Quantum Numbers.* Spatial and temporal spin designations are shown in Figure 39 and correspond to the difference between spatial and temporal translational symmetries introduced in Chapter 3.

TRANSLATIONAL QUANTUM NUMBERS: *Orbital and Magnetic Quantum Numbers*

Figure 39 provides a dimensional symmetry interpretation of quantum states. The modulus-9 representation of the spatial and temporal translational proportions of the first order field of Pascal's Triangle is given in the left of the figure with polygonal symbols corresponding to odd dimensional perspectives. To the right, these same odd spatial and temporal translational symmetries are labeled with spectroscopic and with spin notation.

Conventionally the orbital quantum number is used to distinguish between the types of symmetry states that valence electrons are thought to occupy. Each orbital type has a unique symmetry satisfied by a maximum number of spin-paired electrons. This maximum number is conventionally referred to as the magnetic quantum number. These maximum numbers of pairs begin with 1 (s), then 3 (p), then 5 (d), and then 7 (f). Yet every period or row of the periodic table doesn't contain every type of orbital. Instead, at every second row there is a more complex orbital type. By convention there are a total of eight periods: all of them contain s-orbitals, six contain p-orbitals, four contain d-orbitals, and two contain f-orbitals. As stated previously, I've added a 0th period occupied solely by Hydrogen.

As shown in Figure 39, I propose that atomic valence orbitals or sub-shells arise from the odd translational symmetries of the first-order dimensional symmetry field of Pascal's Triangle. This explains why new orbital types evolve every other period: Even numbered states don't participate directly in bonding because they harbor symmetrically satisfied spin-paired electrons integrated during preceding periods. To help understand what drives this, recall from Chapter 4 that the continuous space-time proportion links all odd translational proportions and all even translational proportions together separately, such that all odd dimensions are in continuous proportion with the finite-odd symmetry of the singular origin and all even dimensions are in continuous proportion with the infinite-even symmetry of the void which produces mirroring spatial-temporal state pairs. This observation provides a plausible symmetrical explanation for why electrons are first incorporated in odd symmetric configurations.

Notice that, with this alternative interpretation, the orbital quantum number corresponds with the spatial period in which the new orbital type appears, which also corresponds to the maximum number of spin-paired electrons allowed in said orbitals, namely the magnetic quantum number. Hence, the orbital and magnetic quantum numbers can be replaced with a single *Translational Quantum Number*. However, notice that these orbital/magnetic quantum numbers represent the geometric symmetry of said translational perspectives versus their degree of differentiation. So, for instance, the triangular number ten still only has three new spatial-temporal pairs of electrons that correspond to the symmetry of its vertices versus its internal structure that also harbors electrons integrated in preceding periods.

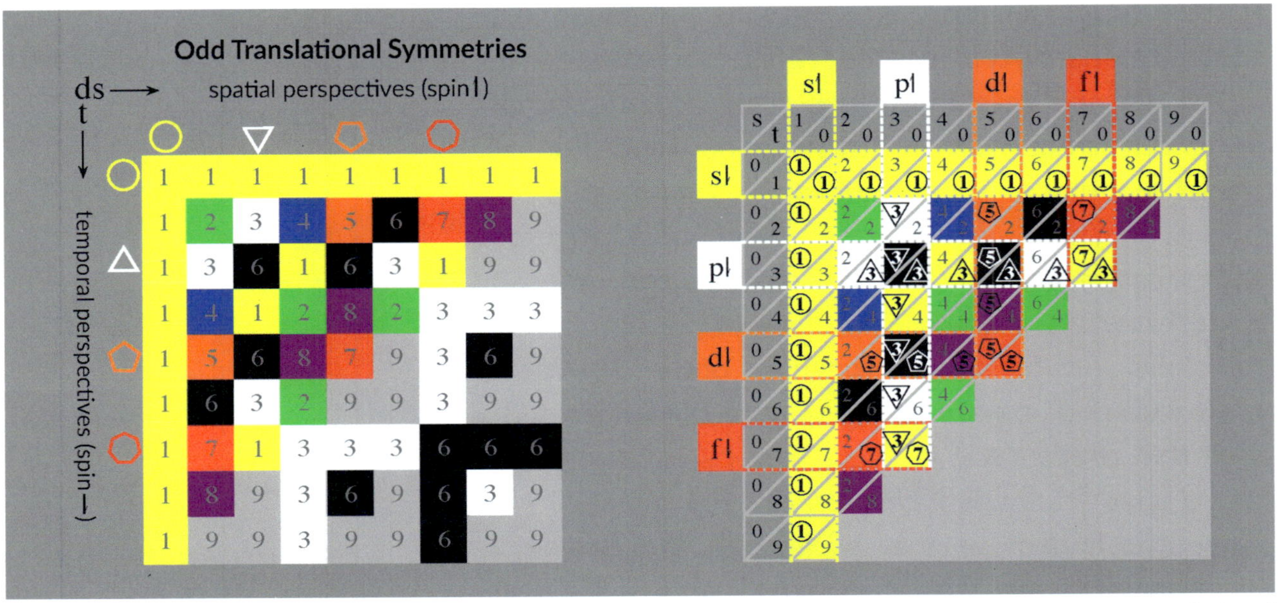

Figure 39. Spatial-Temporal Orientation and Translational Quantum Number: Left – Modulus-9 translational perspectives are shown with polygonal symbols corresponding to odd translational perspectives. Right – Valence shells, using spectroscopic notation, are shown corresponding to odd translational symmetries with spin-up corresponding to spatial orientation and spin-down corresponding to temporal orientation.

The fact that there are in total 8 periods that harbor paired spherical states, 6 periods that harbor paired triangular states, 4 periods that harbor paired *pentangular* states, and 2 periods that harbor paired *heptangular* states can be explained by the modulus-9 analog periodicity of Pascal's Triangle. This is why I elected to name the first 9x9 dimensions the *first order field*. When the first order field is aligned in accordance with incremental spatial constraint imposition, the tertiary phase polarities indicative of the integrated invariance caused by these constraints are pushed out beyond the first nine spatial intervals. This invariant boundary suggests a theoretical limit of 120 atomic elements. Accordingly, the first nine intervals limit the number of each type of sub-shell.

STATE FILLING RULES

The states of one spatial period are generally filled prior to those of the next spatial period. But what governs the filling of states within a given spatial period?

The singular states (s-shell) are always last to be filled within a given spatial period. Furthermore, heptangular states (f-shell) are filled before pentangular states (d-shell); pentangular states are filled before triangular (p-shell), and triangular states before singular states. To understand why this is, examine the governing spatial/temporal constraint interval order in Figure 40, and note that space-filling precedes time-filling consistent with the incremental imposition of spatial and temporal constraints discussed previously.

For a given spatial period the spatially oriented states that occur at the lowest spatial/temporal interval (lowest row) are always those that fill first. This is why sub-shells are filled in descending order. The next lower sub-shell cannot be filled until the entire sub-shell that precedes it is filled, thereby satisfying the paired spatial/temporal symmetry of the sub-shell – a symmetric imperative that supersedes the energetic requirement to satisfy the symmetry states of the next hirer sub-shell. For example examine the filling order of the last period (right-most column): 7/, /7, 5/, /5, 3/, /3, 1/, /1.

Figure 40. Quantum states shown in the governing 2:1 (spatial:temporal) constraint order with spatially oriented states over temporally oriented states (s/t).

THE GENERAL STATE FILLING RULES ARE AS FOLLOWS:

- Spatial intervals are filled from lowest to highest.
- Within each spatial interval, spatial states are filled first from the lowest spatial/temporal interval to the highest.
- Within each spatial interval, temporal states are filled after spatial for a given sub-shell prior to the filling of spatial states in the next lowest spatial/temporal interval.

Note: This filling order is consistent with the well-known Hund's Rules[74]

Using these rules, the periodic symmetry states shown in Figure 40 are transformed into the Periodic Table of Atomic Elements (Figure 41). Wherein, translational symmetry states corresponding to *Translational Quantum Numbers* are shown incrementally satisfied: first by spin-up electrons culminating in half filled sub-shells (spatially oriented translational symmetry states), then by spin-down electrons culminating in fully filled sub-shells (fully satisfied spatial and temporal translational symmetry states). The spatial constraint periods corresponding to *Spatial Interval Quantum Numbers* of Figure 40 are shown relative to the rows of the periodic table.

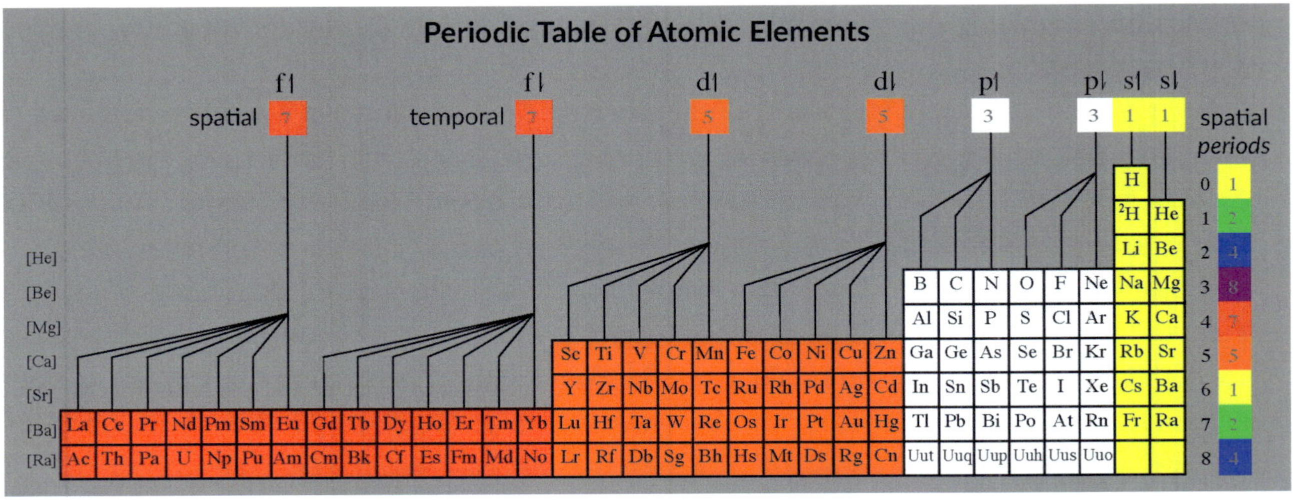

Figure 41. Periodic Table of Atomic Elements derived from the first principals of Dimensional Number Theory. Lines are used to show how electrons add up to ultimately satisfy spatial then temporal translational symmetries.

9.3 SYMMETRY BREAKING

Atomic matter emulates the symmetry of energy imperfectly since dipoles cannot satisfy both spatial and temporal translational symmetries simultaneously. This breaks the symmetry states dipoles occupy within matter into separate spatial and temporal components. The dimensional number symmetry of energy is not restricted in this way since the individual translational point symmetries of Pascal's Triangle have the capacity to satisfy both spatial and temporal orientations simultaneously. For instance, the spatial pentangular number 15 is also the temporal triangular number 15. Dipoles lack this flexibility owning to the ordering of spatial and temporal constraints that defines their chiral cuboctahedral rotational super-symmetry.

This interpretation suggests that translational symmetry breaking affects the stability of atomic energy states and therefore the relative binding energy of electrons in those states. The coming analysis relies on the assumption that electron binding energy is proportional to its affinity for the symmetry of a given state and inversely proportional to its symmetrical freedom within said state. Additionally, the analysis relies on the assumption that the symmetry breaking resulting from disallowed conflicts preferentially influences the electron's symmetric affinity for a given state.

Neutrons appear to act as symmetrical constraints that impose restrictions on the freedom of a given electron-proton pair in order to disallow symmetry conflicts between these alternate spatially and temporally oriented translational energy symmetries. Neutron buildup (i.e. departure from the 1:1:1 ratio of electron:proton:neutron) can be explained by the necessity to disallow multiple symmetry conflicts for a given electron-proton pair perhaps owing to the electron's freedom to transition to energy symmetry states beyond its ground valence state.

Examination of atomic isotope abundances reveals that neutrons are often acquired in pairs. This pairing suggests a symmetric driving force for nuclei to acquire neutrons to disallow conflicts between spin-up (spatial) and spin-down (temporal) states. Analysis of nuclear spin data corroborates this since the spins of paired nucleons appear to cancel.[75] For instance the nucleus of Hydrogen is spin ½ with a single proton, then becomes spin 1 when an additional neutron is acquired (deuteron), then when a second proton is acquired it becomes Helium spin ½, and finally when Helium acquires its second neutron its nucleus becomes spin 0. A similar pattern can be observed to continue as additional nucleons are acquired.

In contrast the coming analysis of first ionization energy (the energy required to remove the outermost electron from a given element) versus atomic number (Figure 42) suggests that all spatial states of a given translational symmetry must be filled before their temporal counterparts are filled – this is consistent with the state filling rules adopted in the last section. These data therefore suggest competition between the symmetric driving force of the nucleus to disallow paired spatial-temporal symmetry conflicts and the symmetric driving force influencing electrons to satisfy the geometric symmetry of translational states.

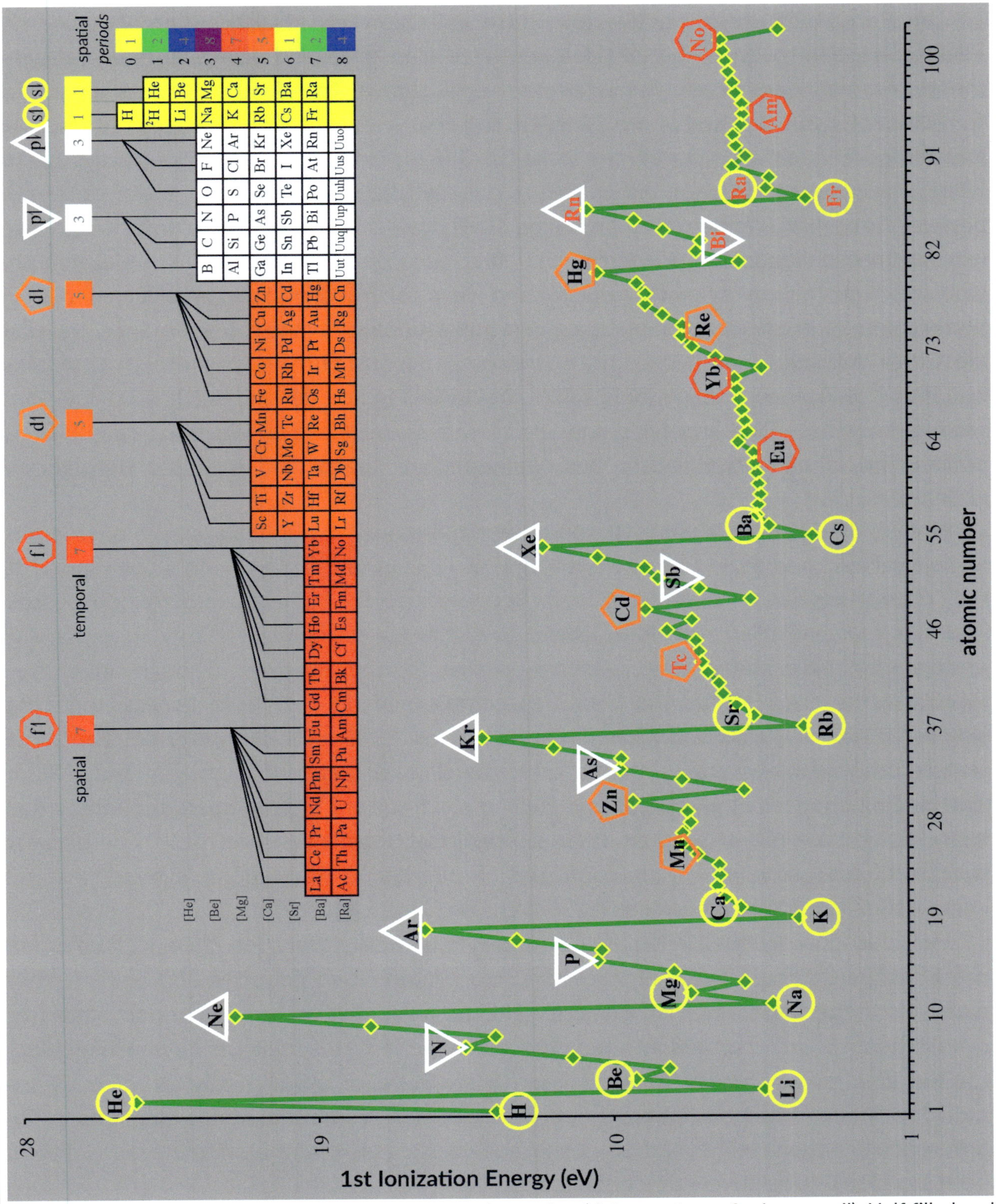

Figure 42. Plot of first ionization energy versus atomic number (data courtesy of science.co.il). Half filled and filled shells are shown with polygon symbols indicative of their translational symmetries.

There are obvious peaks and valleys in the ionization energy distribution of Figure 42: peaks correspond with both half-filled and filled sub-shells, and valleys correspond with the states that immediately follow peaks. Recall that in half-filled sub-shells all spatial translational symmetry states are satisfied and that in fully filled sub-shells both spatial and temporal symmetry states are satisfied. Since binding energy peaks always occur when translational states are either half or fully satisfied, it follows that these states are preferred. A more symmetrically preferred state implies less symmetry conflicts and also less symmetric freedom. An electron that first occupies a given translational symmetry (sub-shell) has the lowest binding energy and the most freedom. Such an electron must in a sense spread itself out to emulate a given translational symmetry. For instance, the first electron in a triangular state can only be triangular if it sustains the symmetry by spreading out in thirds. Hence, the periodic table shown in Figures 41 and 42 has angled lines feeding-in to half-filled and filled sub-shells. This corresponds to electrons cumulatively building up to ultimately satisfy the symmetries of spatial and temporal translational geometries.

Notice that half-filled singular-spherical dimensional states (alkalis) are always the lowest valleys. And that, with the exception of Helium, the highest peaks correspond to fully filled triangular symmetry states, and in every case, the highest peaks are noble gases. The fact that half-filled s-orbitals always have the lowest binding energy suggests that electrons in these states have symmetry similar to free electrons. Though unlike free electrons, the alkali electron is bound to complement its proton. In general then, the singular states of alkalis and alkaline earths are less symmetrically preferred than other translational symmetry states, that is, after the singular states of Hydrogen and Helium. The binding energies of Hydrogen and Helium electrons are high compared to most other elements and are higher than all other s-orbital electrons; I interpret this to be because their electrons are relatively un-conflicted by alternative translational symmetry states compared to other s-orbital electrons.

Per ionization energy, Helium is the most symmetrically preferred state. Helium's ideal paired spatial-temporal symmetry harbors two protons, two neutrons and binds its two electrons tighter than all other elements. Helium forms the first even periodic state (the *dyad*) when it is incorporated into the next spatial period. Helium is the key building block for the ideal symmetrically satisfied internal structure of atoms, as is affirmed by the radiation of alpha particles (2 protons and 2 neutrons) from meta-stable elements.[76] The difference between Helium and other subsequent states is that the symmetry of Helium is both satisfied and relatively un-conflicted.

The stability of Helium's paired protons and neutrons supports the notion that spatial and temporal states are paired in the nucleus similar to the way they are paired for electrons, as in the case of the covalent electrons of the H_2 molecule. The symmetrical forces within the H_2 are *saturated* (satisfied) analogous to the way the nuclear forces on protons and neutrons are thought to behave. This is referred to as the *saturation of nuclear forces.*[77] The states that protons occupy appear to complement those of electrons since together they are satisfied – this suggests that protons have periodic symmetry states like electrons. Modeling proton states after electron states is referred to as the *nuclear-shell model.*[78] This type of model is supported by the temporal/spatial symmetry of the anti-field shown in Figure 38 with the equivalent revisions to the shell model as were proposed for the quantum states of electrons. I will refrain from discussing the notion of *magic numbers* associated with the conventional nuclear-shell model, as I can find no theoretical basis to substantiate them within the context of dimensional number theory.

As discussed previously, the symmetry of energy is flexible because the same state may, for example, be both spatially singular and temporally triangular; or, in another instance, be both spatially pentangular and temporally triangular. The dipole fields of electrons and protons lack this flexibility. As a result, each translational symmetry state must be constrained to either spatial or temporal orientation separately through the acquisition of neutral invariance (neutrons). The neutron then appears to act as the stabilizer or placeholder for a given symmetry state.

The ionization energies of singular s-shell states beyond Helium appear to be suppressed, suggesting that the acquisition of neutrons causes these states to be less symmetrically preferred owing the symmetry breaking required to separate them from their alternate translational orientations. For instance, the alternate temporal triangular orientation of Lithium's singular spatial-spin electron appears to be theoretically disallowed by its third neutron. Going a step further, Lithium is most abundant with a forth neutron, which it acquires to disallow conflicts with the competing singular temporal-spin state ultimately satisfied by Beryllium. But Beryllium must acquire a fifth neutron, presumably to disallow conflicts with the state satisfied by Boron's single spatial-triangular electron (based on this theoretical explanation for the function of neutrons).

The triangular dimensional symmetry appears to suppress the symmetry of other dimensional perspectives since, beyond Helium, its full shells are most tightly bound. One may surmise that the likely explanation for this is that the triangular dimensional symmetry is the most preferred after Helium. As affirmation, consider the first unstable radioactive element Technetium (Tc), which corresponds to the spatial pentangular number 15: It's symmetry conflicts with the temporal triangular number 15 corresponding to the noble gas Xenon. It appears as though this symmetry conflict is the most probable symmetrical explanation for the instability of Technetium.

As Figure 42 shows, the range of binding energies narrows as atomic number increases. Correspondingly, the rate of neutron integration exceeds the arithmetically proportioned integration of protons relative to atomic number. And this neutron buildup corresponds with atomic instabilities that ultimately lead to radioactive decay. Beyond Lead's 126 neutrons, there are no stable isotopes of any element, Bismuth having recently been found to slowly decay.

Based on the proposed function of neutrons, nuclear instabilities appear to be associated with the buildup of neutrons required to disallow conflicting spatial and temporal symmetric orientations. Further, as each neutron is presumed to hold a specific state invariant, this buildup suggests that electron-proton pairs increasingly require that more than one state be held invariant in order to avoid these symmetry conflicts. Given that electrons are regularly excited to states beyond their base valence state, it seems quite likely that the states they regularly visit may also be held invariant by neutrons to resolve symmetry conflicts. This integrated invariance in effect enables the subsequent integration of proton-electrons pairs through beta decay – *nucleosynthesis*.

9.4 Compound States

Based on the preceding analysis of atomic elements, it's reasonable to consider the symmetry states of energy and anti-energy to be like a virtual archetype that matter emulates as it evolves in complexity since the states that electrons occupy in matter emulate the energy symmetry states of Pascal's Triangle. In other words energy governs matter and so matter's symmetry states are sustained by the symmetry of energy. But atomic matter is limited to the periodic table of elements, which, as discussed previously becomes unstable within the first-order analog field due to symmetry conflicts. So if the symmetric states of energy and anti-energy dipoles within matter were to be limited to the states of atomic elements, they would be limited to the first-order field owing to neutron buildup and to the accompanying nuclear instabilities.

However, atomic bonding seems to provide energy and anti-energy dipoles with a means of attaining higher symmetry states without acquiring additional neutrons. These dipoles appear to leverage the lighter elements (elements less encumbered by symmetry conflicts) to evolve organically, thereby circumventing the symmetric limitations of atomic matter while simultaneously taking advantage of the diversity of matter states.

Atomic electrons appear to have the capacity to occupy higher symmetry states without acquiring additional neutrons, primarily through ionic bonding, covalent bonding, and bonding that has a mixture of both characteristics. *Ionic bonding* is generally characterized as a give-take relationship between atoms of highly differing *electronegativity*[*]. So in ionic bonding, for instance, the relatively free singular state of the electron in the Sodium atom may advance to the next spatial interval to satisfy the final triangular symmetry state of Chlorine. By joining with Chlorine the outer Sodium electron effectively leaves its former state for a higher more symmetrically preferred one. *Covalent bonding* is generally characterized as a sharing relationship between atoms with similar electronegativity. In covalent bonding electrons appear to satisfy each other's symmetric desire for spatial-temporal state pairing. So, for instance, two Hydrogen atoms prefer to bond covalently to form the H_2 molecule, wherein one electron occupies Hydrogen's singular spatial spin-up state and the second satisfies the dimensional state comparable to Helium's singular temporal spin-down state.

In the middle between these two bonding extremes there are atomic bonds that have both covalent and ionic characteristics, where bonding is a blend of ionic taking and covalent sharing of electrons. This type of bonding creates a secondary weaker form of bonding referred to as *hydrogen bonding*. Hydrogen bonding is quite common and is important for both inorganic molecules such as H_2O and for organic molecules such as DNA. When Hydrogen is bonded to a highly electronegative atom such as Oxygen or Nitrogen, both of which seek to satisfy their highly preferred triangular p-orbital symmetry states, the bonding is mostly covalent since Hydrogen's electron symmetry state is almost as preferred as the states it's electron occupies when bonding with either Nitrogen or Oxygen. A direct comparison of ionization energies in Figure 42 affirms this interpretation.[†] In this situation the p-orbital states are slightly more preferred, so electron sharing is not quite equal. This leaves Hydrogen's proton less than fully satisfied, and so it has a weak residual ionic charge that is proportional to the amount of ionic taking in which its covalent partner is engaged in. This residual ionic charge creates attraction and bonding to other molecules (*intermolecular hydrogen bonds*) and bonding to other parts of the same molecule (*intramolecular hydrogen bonds*).

Yet another means of advancement that dipoles appear to leverage is referred to as hybridization where, for instance, Carbon's four outer electrons (2 singular and 2 triangular) appear to combine to occupy half-satisfied tetrahedral states. Then in one instance of this tetrahedral Carbon geometry, four Hydrogen atoms share their electrons with the single Carbon atom to satisfy its hybrid tetrahedral states, and thereby

---

[*]Electronegativity is a semi-empirical method of estimating bond type developed by Pauling. It is roughly proportional to the sum of the electron affinity (energy to add an electron) and ionization potential (energy to remove an electron).[82]

[†]Oxygen is the next data point after nitrogen in Figure 42.

Hydrogen's electrons are promoted from singular to tetrahedral symmetry states of methane.

Metallic bonding is another type of atomic bonding that is increasingly prevalent as atomic number increases. The primary characteristic of metallic bonding that sets it apart from both covalent and ionic bonding is the delocalization of valence electrons. In a given piece of metal valence electrons are relatively freely shared between all of its atoms, like a *sea of electrons*. Metallic electrons are loosely bound, which is characteristic of dipoles in less preferred symmetry states. However, since they are bound to other atoms of the same or similar disposition, they are able to share their electrons like covalent bonding, but in a delocalized manner that differs from covalent bonding. The fact that atomic elements become increasingly metallic suggests that the symmetry states dipoles occupy in atomic matter are increasingly neutralized by the integration of invariances (neutrons) required to disallow symmetry conflicts. So metallically bound electrons tend to behave like free electrons versus those bound to a specific atom-localized state, since the symmetric influence of those localized states are largely neutralized.

However, given that the evidence for a continuum of states is the same evidence that supports the existence of these dipoles, there is no reason to think that these dipoles themselves are inherently limited to the first order field except when they are within the confines of a given bonding configuration. Rather, the dipoles themselves are formed by a 24-fold analog super-symmetry that extends well beyond the confines of the first order field. So I think the rich diversity symmetry states in inorganic and organic compounds suggest there is an alternate avenue that enables these dipoles to occupy higher, more dimensionally complex, states within the continuum.

Therefore, it's curious to note that atomic bonding ultimately leads to the formation of complex organic molecules such as DNA. The two strands of the DNA double helix are anti-parallel and complementary. This symmetric feature seems quite similar to the distinction between spatial and temporal dimensions. Bonding between these two strands only occurs between specific *complementary base pairs*. The fact that DNA is composed of just four bases adenine (A), thymine (T), cytosine (C), and guanine (G) arranged in reverse order along these anti-parallel strands is very similar to complementary numeric dipole symmetry. So if I consider the complementary base pairs to be like dipoles (A-T and C-G), then their specific affinity to one another (their polarity) is linked to the pairing of their hydrogen bond strengths, which is a specific selectivity that results from the atomic bonding unique to each base. These hydrogen bonds dictate the anti-parallel orientation of these base pairs. In the case of the A-T base pair there are two hydrogen bonds that define their anti-parallel orientation: one between hydrogen and oxygen and one between hydrogen and nitrogen. In the case of the C-G base pair there are three hydrogen bonds that define their anti-parallel orientation: two between hydrogen and oxygen and one between hydrogen and nitrogen. By bonding in this fashion the residual ionic charges (polarities) of base pairs is effectively neutralized. So, by assembling into complex complementary organic dipoles, it appears as though field and anti-field dipoles are able to circumvent the symmetric limitations of inorganic matter and attain higher dimensional states within the continuum – that is of course if the preceding assumptions are correct. The parallels between complementary dimensional number symmetry and DNA are suggestive of a cause-effect relationship that is difficult to ignore, but a firm connection between the two remains to be established.

9.5 COSMOLOGIC IMPLICATIONS OF DIMENSIONAL NUMBER THEORY

THE COSMOLOGIC EVOLUTION OF MATTER

Let me be the first to state that I make no claim to be an expert on the cosmologic processes studied by astrophysicists. However, given that I am presently discussing the possible physical implications of dimensional number theory, it is difficult to avoid crossing over into this less familiar territory. The spatial-temporal symmetries within number are comfortably precise, but symmetry breaking within matter blurs observation of their influence. Never the less I will continue to observe the imperfect from the perspective of the perfect.

The formation of atomic nuclei begins with Hydrogen. Hydrogen occurs most abundantly without neutrons. It is also stable with one neutron, though far less abundant, and is unstable with two neutrons, in which case it decays to Helium. Because it is the sole element in the first spatial constraint interval (period 0), Hydrogen seemingly has no need of a neutron. It appears to be constrained much in the same way as singular translational perspectives are constrained, that is, only by inversion and therefore without neutral numeric invariance. Yet, together with Helium, Hydrogen is also present in the second spatial constraint interval (period 1). When paired with Helium in this spatial constraint interval, Hydrogen must acquire a neutron (Deuterium) such that the state of its electron-proton pair may be differentiated symmetrically from those of Helium's additional pair. However, Hydrogen's electron-proton pair may easily be excited further so as to occupy the spin-down state of Helium's singular temporal perspective which causes it to acquire a second neutron (Tritium), thereby paving the way for the evolution of Helium-3.[80] This example suggests that neutrons occupy symmetry states prior to their permanent occupation by electron-proton pairs, a condition consistent with the chronology of neutron build-up followed by fusional decay.

Conventional theory often referred to as *Big Bang Nucleosynthesis* uses the present-day relative abundances of hydrogen (74%) and helium (26%) as the basis of the theory and attempts to explain what must have transpired to achieve these percentages. It holds that protons and neutrons were formed in equal abundance from the primordial quark-gluon plasma of the big bang since the temperature was high enough for them to exchange back and forth in a relative equilibrium. But then as the universe cooled there reached a temperature where neutrons became less stable and their relative abundance began to diminish until the temperature at which deuterium became stable. At this temperature the remaining neutrons combined to form deuterium then helium halting their decay. It is thought that no heavier elements formed until the universe cooled sufficiently for hydrogen and helium condensed into massive stars where heavier elements were then formed through a fusion process involving the capture of neutrons, or through a fission process that involves the breakup/decay of heavier elements.

There are several known issues with the standard Big Bang model, which is predicated on observations of the expanding universe, the relative abundance of hydrogen-helium, and cosmic background radiation. Most interesting to me is that according to Trefil "There is no known process that can change the net particle number of the universe ... by the time the universe is a millisecond old, the balance between matter and antimatter is fixed forever."[81] I've proposed a process whereby both matter and antimatter particles are formed from energy and anti-energy respectively. Moreover I've postulated that the proton is actually a product of the anti-energy field suggesting that there is in fact a perfect complementary balance between matter and antimatter in the universe and that what we consider to be matter is actually a combination of both. The same model also gives cause for our collective misinterpretation given the natural spatial/temporal orientation of the universe, which inverts our perspective of the temporal/spatial anti-field.

I challenge the standard model for the following reasons: First, the standard Big Bang model suggests some sort of chaotic explosion of a superheated quark-gluon plasma from which all the matter in the universe is produced in a fraction of a second, rather than envisioning an arguably more plausible ongoing process whereby atomic particles are formed from dimensional states of energy and anti-energy – a model consistent with the expanding universe evolving from a singular spatial/temporal energy state. Second, the concept of a superheated quark-gluon plasma is an impossible scenario according to dimension number theory, since the theory suggests these observed *particles* are degenerate symmetries produced when stable cuboctahedral dipoles are perturbed during particle accelerator collisions. Third, temperature is a measure of the kinetic excitation of matter as expressed by the Boltzmann relationship ($E=k_BT$), but neither matter nor its vibrations can be sustained by a singular state since a singular state implies a singular spherical waveform of energy, which by definition can sustain only one wave state and therefore excludes phonon vibrations (heat). It is a fallacy to equate energy with temperature at a time prior to the formation of matter. The very nature of matter requires more than one energy state, which precludes its existence at the time of the singularity.

The theory that hydrogen and helium were formed at the early stages of the universe and that heavier elements formed within stars is not inconsistent with dimensional number theory. As manifestations of the first and second spatially oriented intervals of the energy and anti-energy field, hydrogen and helium apparently did and still *are* forming there. It is the notion of linear time and our matter-based perspectives that limit conventional theories because we try to compact all that is at present into a singularity that existed only in the past rather than contemplating an early universe composed entirely of energy wherefrom matter evolved and is still evolving.

If energy and anti-energy fill a continuum of spatial-temporal states, as suggested by the dimensional interpretation of Pascal's Triangle, all spatial and temporal states must be ever-present rather than fleeting historical conditions. How else could the complementary dimensional super-symmetry of cuboctahedral dipoles be sustained? And likewise, how could the virtual periodic symmetry states of atomic elements continue to exist unless all spatial-temporal states exist concurrently as pre-determined by their continuous proportions?

The continued creation of matter from energy through the consumption of void provides a plausible explanation for the continued expansion of the universe – much more palatable than the present theories of *dark matter* and *dark energy* intended to explain this phenomenon. The assumption that all matter in the universe arose from an initial *big bang* needs some serious reconsideration in my opinion.

SYMMETRIC SOURCES OF MASS AND GRAVITY?

Physicists equate energy and mass through Einstein's famous formula $E=mC^2$ such that if either the mass or the energy of a thing is known, then the other may be known, which means, for instance, that the mass of the electron is indirectly determined by its observed energy. The only directly measurable and by far the most massive components of atomic elements are protons and neutrons whose masses are essentially the same, since a neutron is the sum of an electron and proton. So essentially all the mass in the universe may be attributed to protons, if one excludes mass inferred by energy.

Dimensional number theory suggests that protons arise from the spatial/temporally constrained integral perspective of anti-energy (a 1-8 temporal-spatial cuboctahedral dipole field oriented to its spatial pole). And, since protons arise from the singular temporal/spatial perspective of the anti-field and from its invariance relative to the spatial/temporal field, they belong to an inverted perspective and naturally compose the nucleus of an atom. Numeric symmetry also suggests that electrons are spatially oriented spatial-temporal dipoles that arise to the spatial/temporal cuboctahedral dipole field of energy. And the electron is attracted to the proton presumably because they are complements and because both are also poles of the spatial cubic dipole field. Thus the proton restricts the freedom of the electron even though it is separated from it by a neutral void, much as someone weighed down by a ball and chain – the drag of the anti-field on the field.

Then why is it that protons and electrons differ so dramatically in mass, if they are just opposite spatial poles? Based on dimensional number theory, what else can mass be but a manifestation of the anti-field's natural resistance to being constrained to the perspective of the field? If one accepts the theory that both electron and proton arise from the complementary cuboctahedral dipole super-symmetry, then one is left to assume that their extreme mass difference must be associated with the inverted orientation imposed on the anti-field so that it conforms to the perspective of the field, as this is the primary distinction between the two dipoles. And, by extension, it is quite possible that mass could be influenced by other such constraints that oppose natural orientations, as with the case of the electron in a spin-down state. Therefore, the theoretical energy-mass relationship appears to be related with the energy required to bind anti-energy dipoles to the spatial/temporal orientation of the energy field.

Since numeric symmetry suggests that mass arises from anti-energy field dipoles that are bound to an energy field perspective, what does it suggest about gravity? Gravity is known to be a mono-polar force between all masses, and is known to act across vast distances. Its strength diminishes by the square of the distance between said masses because its strength is radially dependent. Further, gravity acts at right angle to the direction of the motion of a given mass such that masses orbit one another under the correct conditions. But in general gravity is a force that is singularly bent on compacting all mass into the smallest possible size – into a star or a black hole for example.

First, there is only one monopole in dimensional number symmetry, and it arises from the constraint of void; and the first constraint of void is the constraint of inversion.

Second, dimensional number symmetry implies that proton mass may be primarily attributed to the binding of the temporal/spatial anti-energy field to the spatial/temporal orientation of the field. Therefore, dimensional number theory suggests that gravity is associated with the mono-polar force of inversion between spatial/temporal and temporal/spatial perspectives, which acts as a mono-polar attractor between all spatially oriented anti-field dipoles (protons) – in a sense beckoning them to return to their common singular dimensional origin and invert to the temporal/spatial perspective.

CHAPTER CITATIONS

[66] Plato trans. Kalkavage, Peter 2001 *Timaeus* Focus Publishing (USA) p 83
[67] Fairbanks, Arthur 1898 *The First Philosophers of Greece* Charles Scribners (New York) p 147
[68] Nakamura et al. (Particle Data Group), JP G 37, 075021 (2010) and 2011 partial update for the 2012 edition (URL: http://pdg.lbl.gov)
[69] Adair, R. 1987 *The Great Design – Particles, Fields, and Creation* Oxford University Press (NewYork) p 289
[70] Fairbanks, Arthur 1898 *The First Philosophers of Greece* Charles Scribners (New York) p 136
[71] Guthrie, K.S. 1988 *The Pythagorean Sourcebook and Library* Phanes Press (USA) p 24
[72] Herzberg, Gerhard 1944 *Atomic Spectra and Atomic Structure* Dover Publications (USA) p 71
[73] Kingery et al. 1976 *Introduction to Ceramics*, 2nd Edition, John Wiley & Sons, Inc. (USA) p 36
[74] Nave, R http://hyperphysics.phy-astr.gsu.edu/hbase/atomic/hund.html#c1
[75] Fuller, Gladys H. 1976 *Nuclear Spins and Moments* J. Phys. Chem. Ref. Data, Vol. 5, No. 4, p.845
[76] Segre, Emilio 1964 *Nuclei and Particles* W.A. Benjamin, Inc. (New York) pp 273-286
[77] Kaplan, Irving 1964 *Nuclear Physics*, 2nd Edition, Addison-Wesley, Inc. (USA) p 509
[78] Kaplan, Irving 1964 *Nuclear Physics*, 2nd Edition, Addison-Wesley, Inc. (USA) p 535
[79] Kingery et al. 1976 *Introduction to Ceramics*, 2nd Edition, John Wiley & Sons, Inc. (USA) p 39
[80] Segre, Emilio 1964 *Nuclei and Particles* W.A. Benjamin, Inc. (New York) p 562
[81] Trefil, James A. 1983 *The Moment of* Creation, MacMillan (USA) p 38

Secrets of Creation

> ... there is no apparent reason why one number is prime and another not. To the contrary, upon looking at these numbers one has the feeling of being in the presence of one of the inexplicable secrets of creation.
>
> D. Zagier[82]

So far I have explored the complementary dimensional symmetry of continuous proportions that are virtual in the sense that these continuous proportions are composed of super-positioned point symmetries. These virtual dimensions produce point-charge dipoles such as the electron and proton that form atomic matter. But, as I noted in the last chapter, there is an *unseen hand*, another related form of dimensional symmetry that effectively triangulates between the symmetry states these dipoles occupy. This dimensional symmetry can be observed experimentally when an atomic electron gains momentum by absorbing a photon (*force particle*) and jumps to a higher energy state, and again when this electron looses momentum by emitting a photon and drops down to a lower state. Unlike the former super-positioned virtual symmetry, this form of geometric symmetry involves changes in physical position in both time and space. Take for instance the integer radial state transitions in the Hydrogen atom, where all models suggest that an electron transitions from one spherical state to another. These state transitions involve photons with specific wavelength/frequency. As I've stated, I don't believe the harmonic oscillator model is sufficiently robust to explain what truly governs these state transitions. So I'm interested in exploring alternative dimensional symmetries that have the necessary attributes to mediate such state transitions.

The electromagnetic field has an element of constraint that causes the direction of the electric field to be orthogonal to the direction of the magnetic field. So for instance, an electron traveling in a straight line generates a magnetic field perpendicular to its own direction. In addition, electromagnetic field symmetry has an element that is similar to a discrete resonant frequency of a string of fixed length. This characteristic symmetry element causes electromagnetic frequencies to be discrete or indivisible (*quantized* — an observation that led science to opt for harmonic oscillator and *string theory* models).

Further, given the symmetric rainbow color dispersion of refracted white light, one can reasonably expect some form of complementary symmetry to accompany the first principles that govern the electromagnetic spectrum. And because the symmetry of the electron both absorbs and emits photons thereby gaining and losing momentum, it is also likely that the symmetry that governs electromagnetism couples with spatial-temporal and temporal-spatial cuboctahedral field dipoles. Additionally, in the theories of quantum mechanics the photon is characterized as a force particle that is its own anti-particle. These characteristics provide vital clues to unravel the geometric mystery of electromagnetism. Each characteristic points to a specific geometric quality or symmetry element. The combination of these characteristics suggests a form of dimensional symmetry that embodies all of the subordinate elements simultaneously.

The symmetry element that produces indivisible quantities in the form of wavelength or frequency is like the symmetry of prime numbers in the sense that they too are indivisible. The symmetry of prime numbers is linked to the triangular symmetry of Pythagorean Triples, which are right triangles with integer-length sides beginning with the 3-4-5 triangle. Hence the symmetry of primes contains the orthogonal symmetry element, which makes it a very interesting avenue to explore as a source of electromagnetic dimensional symmetry.

Proclus attributed the following equations for *near-isosceles triples* to Pythagoras,[83] where *a* and *b* are orthogonal sides, and *c* is the hypotenuse of a right triangle whose hypotenuse and long side differ by one, and *n*, is some defining integer:

$$a = 2n + 1$$
$$b = 2n^2 + 2n$$
$$c = 2n^2 + 2n + 1$$

Near-isosceles triples have a particular triangular geometry that causes *c* and *b* to always differ by one; this is evident from equations for *c* and *b* above. Notice that as *a* and *c* are both multiples of two plus one, both will always be odd, and similarly *b* will always be even, as defined. Also notice that for arithmetically ordered *n*, *a* is also so ordered, but not *b* and *c*. Geometrically *a* and *b* represent orthogonal sides of a given right triangle. Given that *a* is in continuous arithmetic proportion (a linear perspective), it follows from the symmetry of translational dimensional perspectives that *b* should be associated with a triangular perspective (continuous triangular proportion) since its proportion is at right angle to the linear proportion of *a*. These observations help frame the continuous proportions of Pythagorean triples within the spatial-temporal context of the dimensional number field.

Rather than having a set of equations that only work for one class of triples, it is desirable to have a set of equations that governs all triples. The above equations suggest that there are classes of triples ordered by the same geometry that causes the hypotenuse and the long side to differ by one in the case of near isosceles triples. So in order to develop a comprehensive set of equations it is critical to know what geometry causes this constant difference. To accomplish this the first step is to put the above equations in a geometric context, and to do this understanding what the *n* in Pythagoras' equations signifies is essential.

10.1 Geometric Relationships Between Right Triangles and Circles

To understand right triangles the circular symmetry that governs them has to be considered, for every Pythagorean Triple contains an *in-circle* of integer radius and is bounded by *ex-circles* of integer radius. A first step, then, to understanding triples is understanding their defining circles. Consider the defining circles of the 3-4-5 triangle with Figure 43.

To order the analysis of right triangles one must first fix their orientation so that the geometry of their continuous proportions can be related to the continuous spatial and temporal translational proportions of Pascal's Triangle. To do this I'll use *v* for vertical side, *h* for horizontal side, and *d* for diagonal side instead of the conventional signage, and will use the same subscripts for their respective ex-circle radii. No subscript is necessary for the in-circle. I'll show that the continuous proportions that emanate from the in-circle of the 3-4-5 triangle correspond directly to the linear and triangular dimensional perspectives of Pascal's Triangle for the near isosceles class of triples. The perspective of the in-circle (the *finite radial point*) is linear with respect to the continuous proportion of vertical side ex-circle radii, and its orthogonal perspective is triangular with respect to the continuous proportion of horizontal side ex-circle radii.

The 3-4-5 triangle has an in-circle of radius 1, an ex-circle adjacent to its vertical side of radius 2, an ex-circle adjacent to its horizontal side of radius 3. Through geometric proof, Figure 43 shows these three radii are *parts* brought together to form the radius 6 of the ex-circle of the diagonal side: *the first perfect number; for it is counted by its own parts.*[84] The length of the vertical side is equal to the sum of its ex-radius plus the in-radius, and likewise the length of the horizontal side is equal to the sum of its ex-radius plus the in-radius. Also, the length of the diagonal side is equal to the sum of the ex-radii of the vertical and horizontal sides or to the difference between its ex-radius and the in-radius. Further, the area of the right triangle is equal to the product of the vertical and horizontal ex-radii.

Equations 10.1.1 through 10.1.7 detail the set of equalities determined from the geometry of Figure 43, where as defined the colored lines represent lengths equal to respective circle radii:

$$v = r_v + r \qquad (10.1.1)$$

$$h = r_h + r \qquad (10.1.2)$$

$$d = r_d - r \qquad (10.1.3)$$

$$d = r_v + r_h \qquad (10.1.4)$$

$$r_d = r + r_v + r_h \qquad (10.1.5)$$

$$A_{vhd} = r_v r_h \qquad (10.1.6)$$

$$A_{vhd} = r_d r \qquad (10.1.7)$$

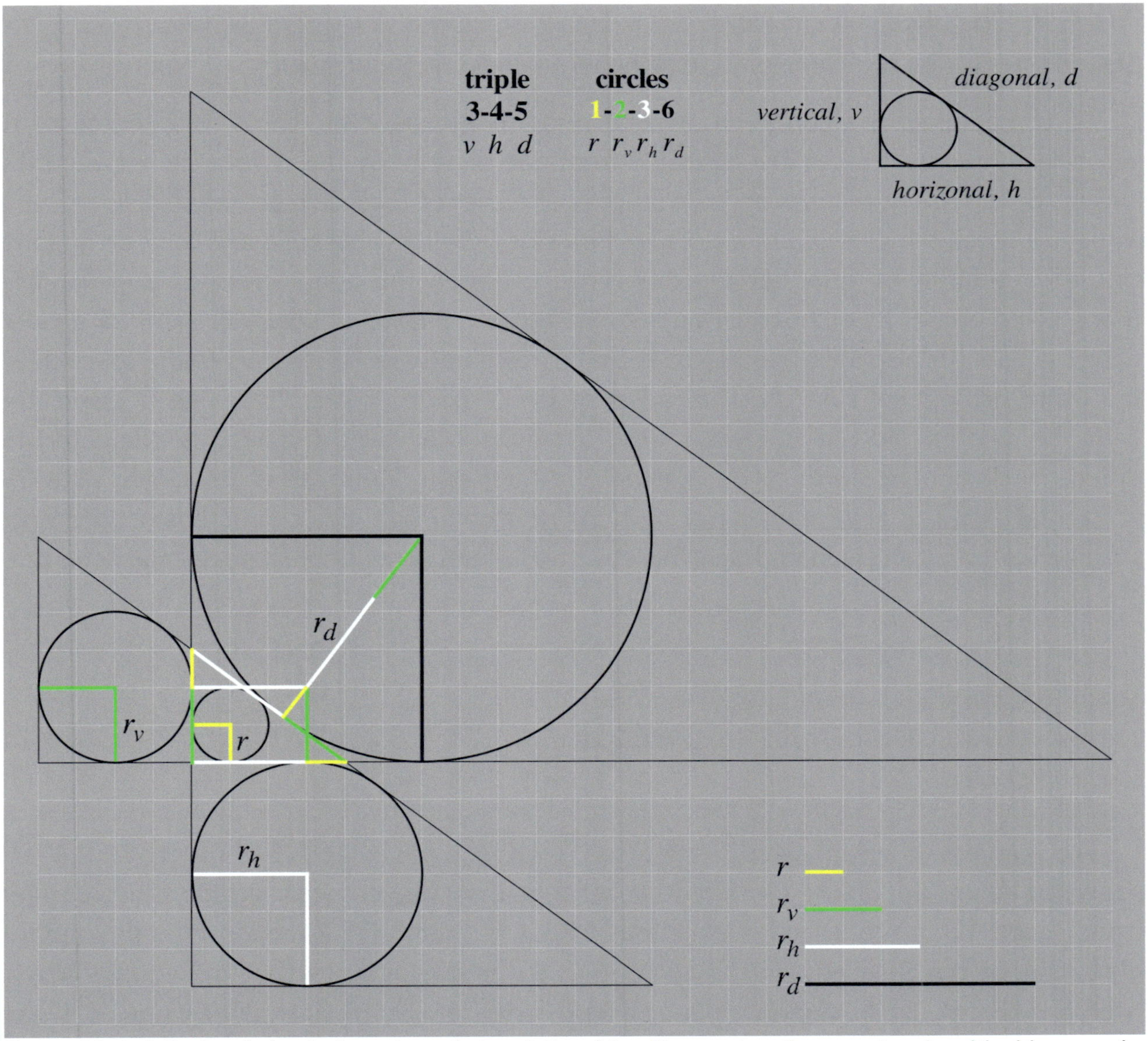

Figure 43. 3-4-5 triangle in-circle and ex-circle relationships: By constructing a rectangle with sides r_v and r_h atop the right triangle the sides of the triangle are found in terms of its in-radius and ex-radii. The radius of the right triangle's hypotenuse ex-circle (r_d) is found to be equal to the sum of the three other radii (r, r_v, r_h). And, the area of the constructed rectangle is found to be equal to the area of the right triangle.

Notice that the ex-radius of the diagonal side is equal to the area of the right triangle divided by its in-radius, a fact that provides a compelling interpretation for the source of perfect numbers since they are the sums (*counted by*) of their divisors (*parts*). Consequently, it should come as no surprise that the set of all *perfect numbers* (6, 28, 496, 8128, 33550336, ...)[85] is contained within the series of r_d's corresponding to near isosceles triples, nor that their associated *Mersenne primes* (3, 7, 31, 127, 8191, ...)[86] correspond to the lengths of the vertical sides of the same triangles since the length v corresponds to the nth arithmetic number, which when summed from 1 to n form the triangular number r_d.

10.2 Pythagorean Triples Classified by their Delta Modula

To characterize the continuous proportions of Pythagorean Triples it's necessary to separate them into classes distinguished from one another by the difference between the lengths of their hypotenuses and the lengths of either of their two orthogonal sides – just as near-isosceles triples are in a unique class since their hypotenuse and long orthogonal side always differ by one. Let the difference between the hypotenuse of a given right triangle and one of its orthogonal sides be designated a *delta modulus* (δ), then, for a given right triangle, δ is also always equal to the difference between the radius of the ex-circle of the other orthogonal side and the radius of the in-circle. If the difference is associated with the vertical side, it is always odd just as is the vertical side; this characteristic difference shall henceforth be designated δ_o. And if said difference is associated with the horizontal side, it is always even just as is the horizontal side; this characteristic difference shall henceforth be designated δ_e. Whereby, the following relationships are established:

Odd Delta		Even Delta	
$\delta_o = d - h$	(10.2.1)	$\delta_e = d - v$	(10.2.3)
$\delta_o = r_v - r$	(10.2.2)	$\delta_e = r_h - r$	(10.2.4)

The geometric relationships expressed with equations 10.2.1 through 10.2.4 are shown in Figure 44. Figure 44 also demonstrates that the geometric relationships derived from Figure 43 and captured in equations 10.1.1 through 10.1.7 remain valid beyond the ideal case of the 3-4-5 triangle. Additionally, the reader will see that the vertical height from the top of the in-circle to the adjacent vertex of the triangle always equals δ_o and that the horizontal length from the side of the in-circle to the adjacent vertex of the triangle always equals δ_e.

The next triangle beyond the 3-4-5 triangle in the continuous proportion of right triangles with $\delta_o = 1$ is the 5-12-13 right triangle (Figure 44). By comparing Figures 43 and 44 it is evident that while δ_o is held constant (fixed) at a magnitude of 1, δ_e increases from 2 to 8 as *r* increases from 1 to 2. The next triangle beyond the 3-4-5 triangle in the continuous proportion of right triangles with $\delta_e = 2$ is the 15-8-17 right triangle (Figure 45). By comparing Figures 43 and 45 it is evident that while δ_e is fixed at a magnitude of 2, δ_o increases from 1 to 9 as *r* increases from 1 to 3. Note also that the geometric relationships established with Figures 43 and 44 remain valid for Figure 45.

The preceding analysis suggest that it is possible to define a given Pythagorean triple triangle and all of its related lengths by its in-circle plus either one of its two ex-circles, which may be described by either δ_o *or* δ_e since they describe the difference between the in-circle and either ex-circle. The analysis also suggests that the continuous proportion of a given fixed-delta class of triples is ordered by integer length of in-radius, as with Pythagoras's equation for near isosceles triples where the integer *n* is given physical context as *r*. Derived, then, from careful study of the proportional relationships of Pythagorean Triples and their circles are the following geometric relationships governed by *r* and either δ_o *or* δ_e:

FOR A GIVEN δ_o...		FOR A GIVEN δ_e...	
$v = 2r + \delta_o$	(10.2.5)	$v = \dfrac{2r^2}{\delta_e} + 2r$	(10.2.11)
$h = \dfrac{2r^2}{\delta_o} + 2r$	(10.2.6)	$h = 2r + \delta_e$	(10.2.12)
$d = \dfrac{2r^2}{\delta_o} + 2r + \delta_o$	(10.2.7)	$d = \dfrac{2r^2}{\delta_e} + 2r + \delta_e$	(10.2.13)
$r_v = r + \delta_o$	(10.2.8)	$r_v = \dfrac{2r^2}{\delta_e} + r$	(10.2.14)
$r_h = \dfrac{2r^2}{\delta_o} + r$	(10.2.9)	$r_h = r + \delta_e$	(10.2.15)
$r_d = \dfrac{2r^2}{\delta_o} + 3r + \delta_o$	(10.2.10)	$r_d = \dfrac{2r^2}{\delta_e} + 3r + \delta_e$	(10.2.16)

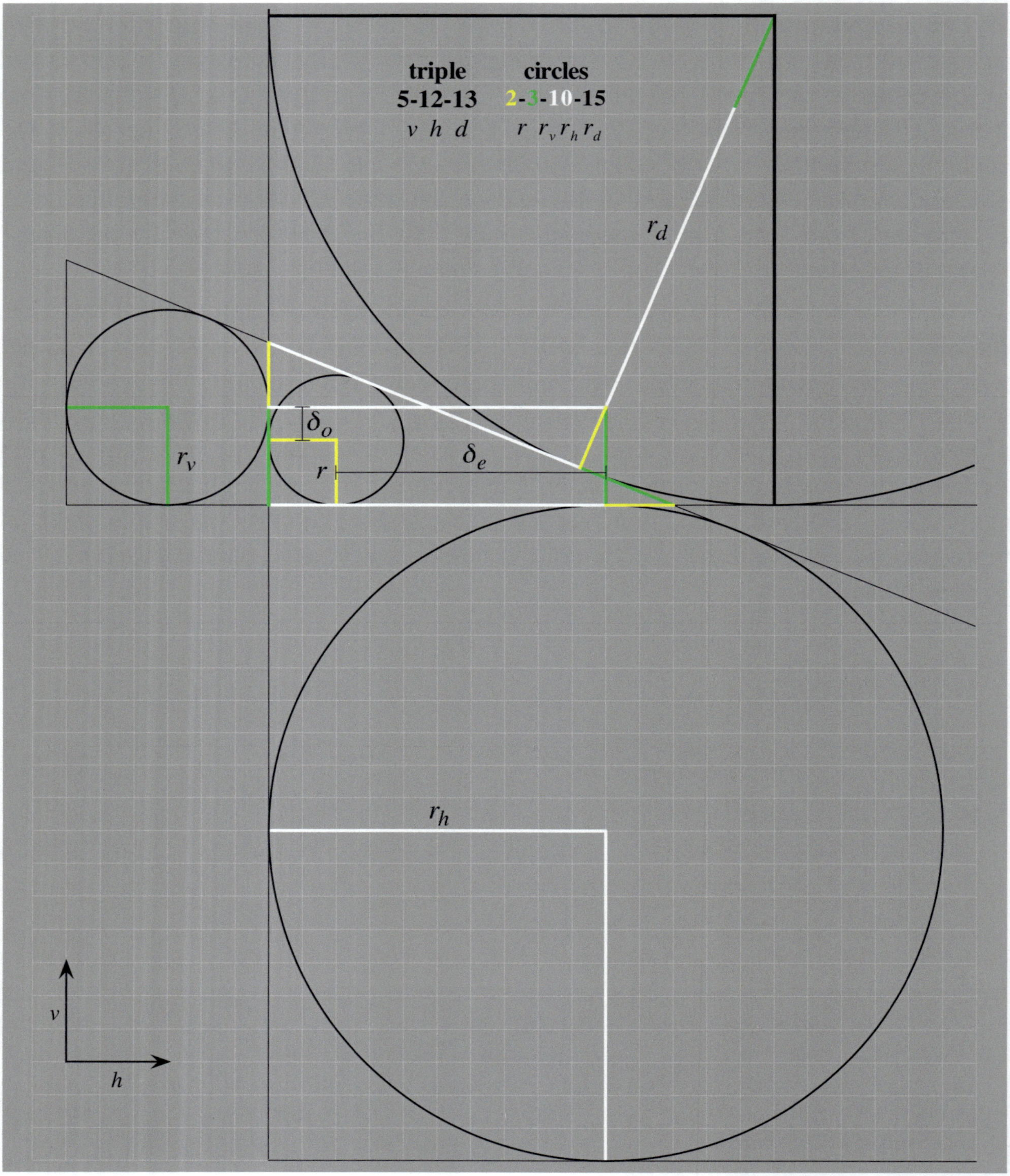

Figure 44. In-circle/ex-circle diagram of the 5-12-13 Pythagorean triple triangle: The first triple triangle beyond the 3-4-5 triangle that has a δ_o = 1 (δ_e increases from 2 to 8). δ_o and δ_e are shown as the difference between ex-circle radii and the in-circle radius. The geometric relationships established with Figure 43 remain valid.

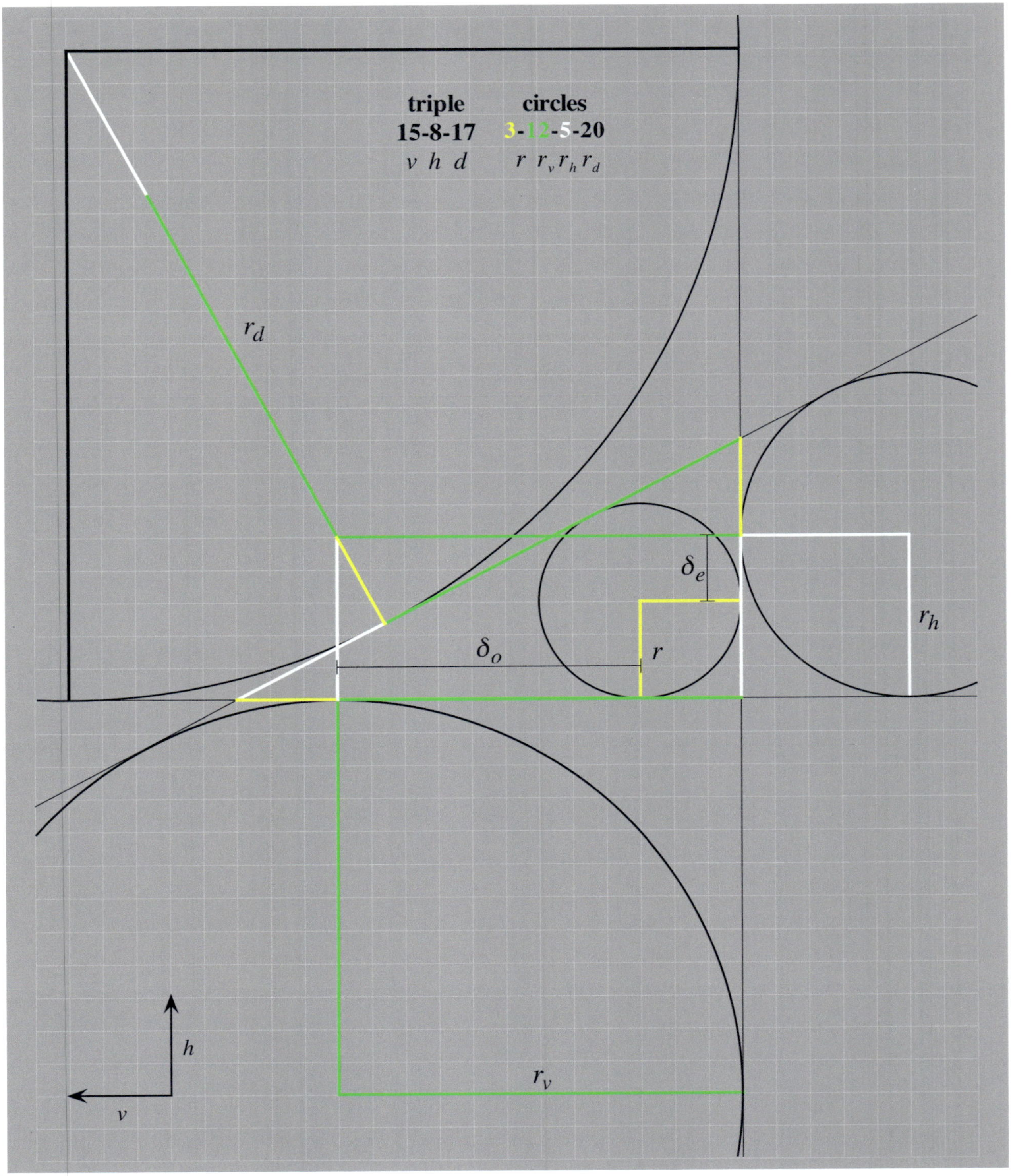

Figure 45. In-circle/ex-circle diagram of the 15-8-17 Pythagorean triple triangle: The first triple triangle beyond the 3-4-5 triangle that has a $\delta_e = 2$ (δ_o increases from 1 to 9). The geometric relationships established with Figures 43 and 44 remain valid. Note: The figure has been rotated by 90° with respect to the previous two figures so that it will fit on the page with the same scale.

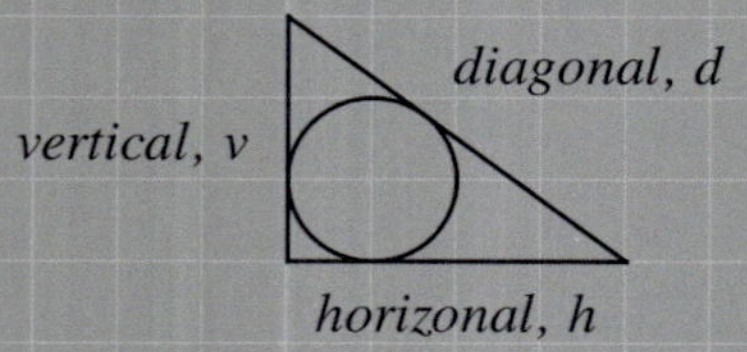

incircle	excircles			triples			differences	
r	r_v	r_h	r_d	v	h	d	δ_o	δ_e
1	2	3	6	3	4	5	1	2
2	3	10	15	5	12	13	1	8
3	4	21	28	7	24	25	1	18
4	5	36	45	9	40	41	1	32
5	6	55	66	11	60	61	1	50
6	7	78	91	13	84	85	1	72
7	8	105	120	15	112	113	1	98
8	9	136	153	17	144	145	1	128
9	10	171	190	19	180	181	1	162
10	11	210	231	21	220	221	1	200
11	12	253	276	23	264	265	1	242
12	13	300	325	25	312	313	1	288
13	14	351	378	27	364	365	1	338
14	15	406	435	29	420	421	1	392
15	16	465	496	31	480	481	1	450
16	17	528	561	33	544	545	1	512
17	18	595	630	35	612	613	1	578
18	19	666	703	37	684	685	1	648
19	20	741	780	39	760	761	1	722
20	21	820	861	41	840	841	1	800
21	22	903	946	43	924	925	1	882
22	23	990	1035	45	1012	1013	1	968
23	24	1081	1128	47	1104	1105	1	1058
24	25	1176	1225	49	1200	1201	1	1152
25	26	1275	1326	51	1300	1301	1	1250
26	27	1378	1431	53	1404	1405	1	1352
27	28	1485	1540	55	1512	1513	1	1458
28	29	1596	1653	57	1624	1625	1	1568

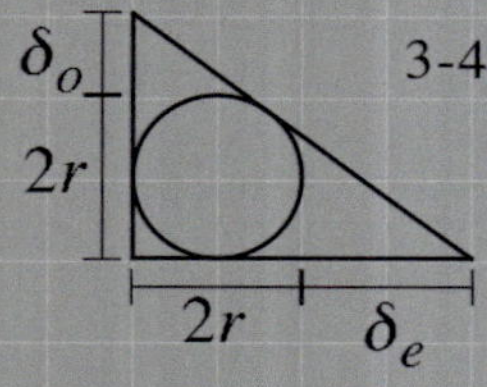

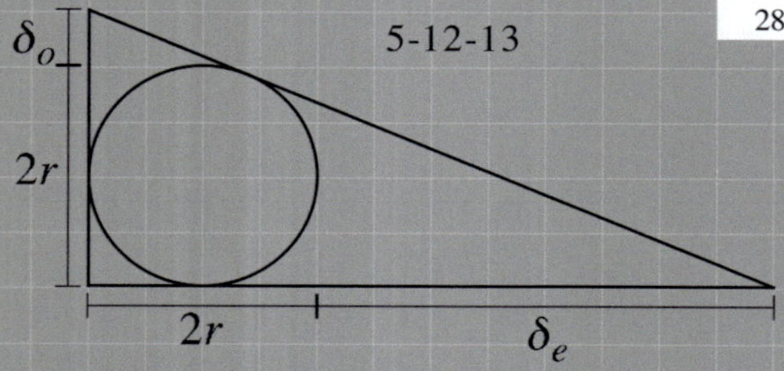

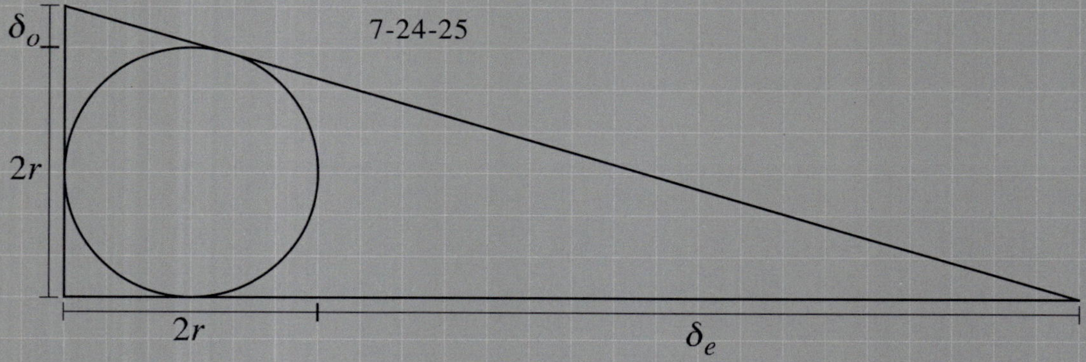

Figure 46. δ_o=1 Pythagorean Triples: The first three triangles of the series are shown (3-4-5, 5-12-13, and 7-24-25). The continuous proportions of component lengths are provided in the accompanying table. δ_o and δ_e are shown equal to the length of their respective sides less the diameter of the in-circle.

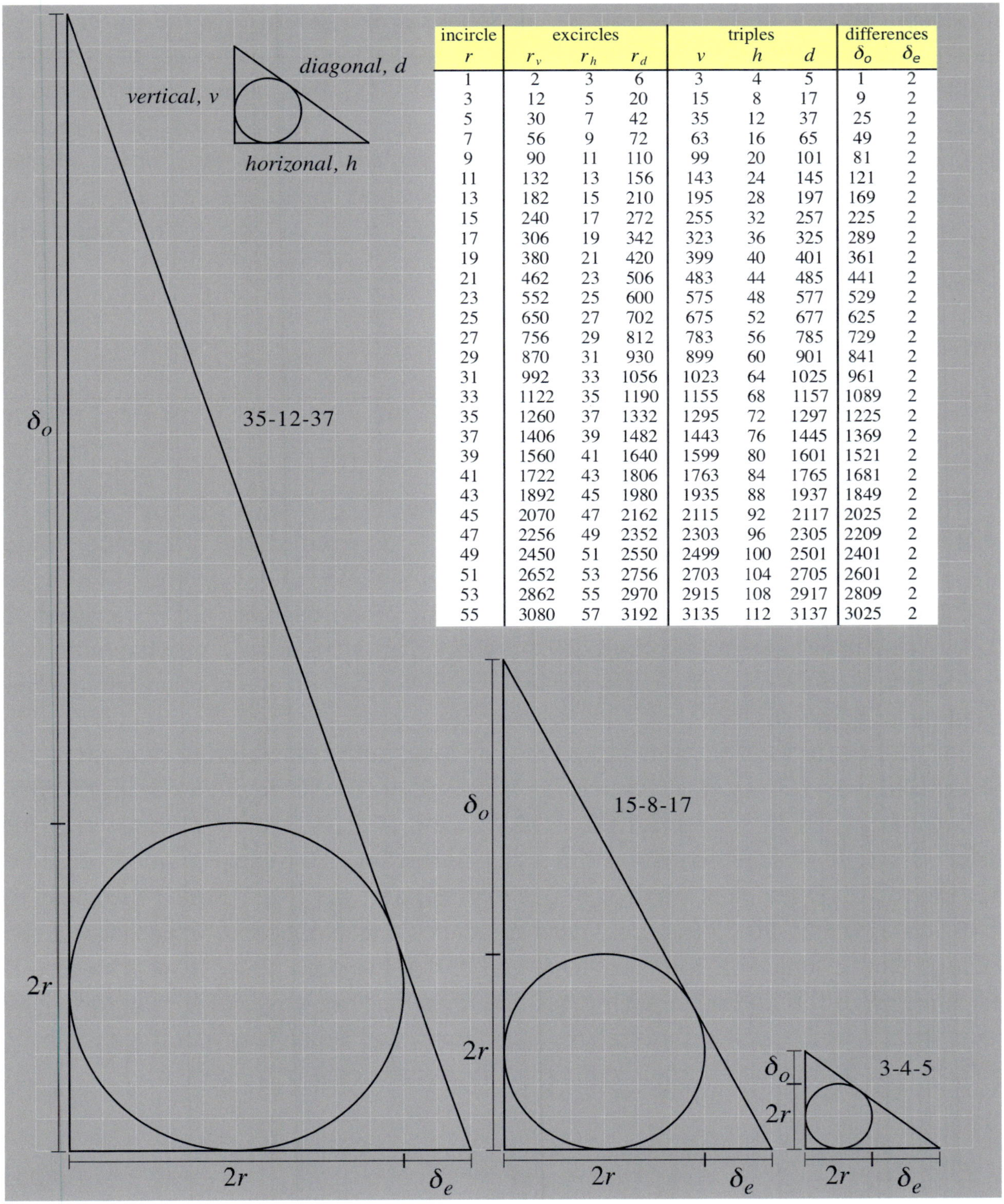

incircle	excircles			triples			differences	
r	r_v	r_h	r_d	v	h	d	δ_o	δ_e
1	2	3	6	3	4	5	1	2
3	12	5	20	15	8	17	9	2
5	30	7	42	35	12	37	25	2
7	56	9	72	63	16	65	49	2
9	90	11	110	99	20	101	81	2
11	132	13	156	143	24	145	121	2
13	182	15	210	195	28	197	169	2
15	240	17	272	255	32	257	225	2
17	306	19	342	323	36	325	289	2
19	380	21	420	399	40	401	361	2
21	462	23	506	483	44	485	441	2
23	552	25	600	575	48	577	529	2
25	650	27	702	675	52	677	625	2
27	756	29	812	783	56	785	729	2
29	870	31	930	899	60	901	841	2
31	992	33	1056	1023	64	1025	961	2
33	1122	35	1190	1155	68	1157	1089	2
35	1260	37	1332	1295	72	1297	1225	2
37	1406	39	1482	1443	76	1445	1369	2
39	1560	41	1640	1599	80	1601	1521	2
41	1722	43	1806	1763	84	1765	1681	2
43	1892	45	1980	1935	88	1937	1849	2
45	2070	47	2162	2115	92	2117	2025	2
47	2256	49	2352	2303	96	2305	2209	2
49	2450	51	2550	2499	100	2501	2401	2
51	2652	53	2756	2703	104	2705	2601	2
53	2862	55	2970	2915	108	2917	2809	2
55	3080	57	3192	3135	112	3137	3025	2

Figure 47. δ_e=2 Pythagorean Triples: The first three triangles of the series are shown (3-4-5, 15-8-17, and 35-12-37). The continuous proportions of component lengths are provided in the accompanying table. δ_o and δ_e are shown equal to the length of their respective sides less the diameter of the in-circle.

Figures 46 and 47 indicate that the continuous proportion of δ_o is the set of all odd squares (1, 9, 25, 49, 81, ...) and the continuous proportion of δ_e is the set of all squares scaled by two (2, 8, 18, 32, 50, ...). When δ_e is fixed, δ_o grows by the square of the in-radius, and when δ_o is fixed, δ_e grows by double the square of the in-radius. The immediate practical application of this observation is that a calculation can be made for the smallest in-circle possible for a given fixed-delta. This smallest in-circle is the origin of the continuous triples proportions of given fixed-delta class. As one might expect, the relationships are different for fixed δ_o and δ_e:

$$r_1 = \delta_o^{\frac{1}{2}} \tag{10.2.17}$$

$$r_1 = \left(\frac{\delta_e}{2}\right)^{\frac{1}{2}} \tag{10.2.18}$$

The continuous proportions of a given fixed-delta class of triples is governed by the continuous proportions of its consecutive in-radii. In the case of δ_o=1 (Figure 46) in-radii are in continuous arithmetic proportion with linear translation of 1, whereas in the case of δ_e=2 (Figure 47) in-radii are in continuous arithmetic proportion with linear translation of 2. This condition suggests that in-radii of a given class increase in proportion with their fixed-delta. However, in both cases the initial in-radius is 1, so any relationship involving fixed-delta is masked.

In order to observe the relationship between fixed-delta and the proportions of in-radii other fixed-delta classes of triples must be analyzed, since in these other classes the initial in-radius is greater than 1 per equations 10.2.17 and 10.2.18. Take for example the fixed-delta class δ_e=8. The first triangle of this variety is the 5-12-13 triple (Figure 44). Based on equation 10.2.18 the initial in-radius of this class has a length of 2. The next triangle in the δ_e=8 class is the 21-20-29 triangle whose in-radius is 6 (see Appendix A for sample triples tables). Notice that the in-radius increased by 4 between these two triangles. By examination of this and other such cases one finds that the length of in-radii for a given fixed-delta increase with arithmetic proportion translated by the length of the given fixed-delta divided by the initial in-radius of a given fixed-delta class of triples:

$$r_{n+1} = r_n + \frac{\delta_{fixed}}{r_1} \tag{10.2.19}$$

Also, the in-radius of any given triple can be calculated if both δ_o and δ_e are known:

$$r_n = \left(\frac{\delta_o \delta_e}{2}\right)^{\frac{1}{2}} \qquad (10.2.20)$$

Using the preceding relationships the continuous proportions of any fixed-delta class of Pythagorean Triples can be calculated with a spreadsheet designed to allow the input of a fixed-delta (Appendix A). Next we'll examine these continuous proportions in terms of the equations discussed in Chapter 4.

10.3 CONTINUOUS TRIANGULAR PROPORTION VARIANTS

For near isosceles triples (δ_o=1, Figure 46) the combination of r_h (3, 10, 21, ...) and r_d (6, 15, 28, ...) continuous proportions corresponds to the continuous triangular proportion of Pascal's Triangle when the two proportions are interleaved (3, 6, 10, 15, 21, 28, ...). And, the continuous proportions of r_v and r series correspond to the continuous arithmetic proportion of Pascal's Triangle. Hence v, the sum of r_v and r, also has continuous arithmetic proportions, but with translations of 2 so as to be composed of the set of all odd integers. The continuous proportions of v and r_d are paired since r_d equals to sum of arithmetically proportioned integers from 1 to v. And r_h equals r_d minus v since it is the triangular number that precedes r_d.

In the case of δ_e=2 (Figure 47) interleaving the r_v and r_d series corresponds to twice that of triangular number series of Pascal's Triangle (2, 6, 12, 20, 30, 42, ...). The r_v series of δ_e=2 (2, 12, 30, ...) corresponds with double the r_d series of δ_o=1, although it begins one term ahead. And, the r_d series of δ_e=2 (6, 20, 42, ...) corresponds with the double r_h series of δ_o=1. The r_h and r series of the δ_e=2 class are arithmetically proportioned, yet with translation of 2 so as to be composed of only odd integers.

Notice then that for any given fixed-delta series of triples r is always in continuous arithmetic proportion, r_d is always in continuous triangular proportion, and r_v and r_h switch between the two depending on whether δ_o or δ_e is fixed (if δ_o is fixed r_v is in continuous arithmetic proportion and r_h is in continuous triangular proportion, if δ_e is fixed r_h is in continuous arithmetic proportion and r_v is in continuous triangular proportion). It appears to be a requirement, stemming from the constraint of orthogonality between the linear and triangular perspectives, that r_v and r_h are at right angles and that they alternately conform to either linear or triangular perspectives. But r_d, though seemingly triangular, is not orthogonal to the linear perspective; rather it is the sum of both linear and triangular proportions, and, in so being, is apparently a hybrid form of both constrained to the diagonal between the two.

Moreover, from the fixed δ_o perspective of a given triangle, r_v appears linear and r_h triangular. Yet from the fixed δ_e perspective of the same triangle, r_v appears triangular and r_h linear, as implied by the exchange of linear and triangular relationships between r_v equations 10.2.8 and 10.2.14 and between r_h equations 10.2.9 and 10.2.15. So r_d might be more accurately conceived of as the synthesis of two dual (linear-triangular) perspectives. And, as the proportions of the sides of these right triangles always conform to the proportions of their corresponding ex-circles, anything observed of the latter may be applied to the former.

How do these observations help us characterize the proportions of the various component lengths of triples? Recall that *the continuous space-time proportion equality* (4.2.1) links the triangular perspective to the singular perspective such that the difference between terms grows arithmetically versus being constant, as is the case for arithmetic proportion.

$$^{s}m_{t+1} = 2\,^{s}m_{t} - {}^{s}m_{t-1} + {}^{s\text{-}2}m_{t+1} \tag{4.2.1}$$

In the specific case of the continuous triangular proportion of Pascal's Triangle (s = 3), the $^{s\text{-}2}m_{t+1}$ term always equals 1, since the singular perspective it corresponds to has continuous equal proportion. However, in a more generalized form of continuous triangular proportion this term could be continuously equal to integers other than 1. To illustrate, let $^{s\text{-}2}m_{t+1} = i$, where *i* may be considered a certain increment of triangular amplification that increases the difference between terms of a given proportion with linear translations of *i*. The *continuous amplified triangular proportion equality* is then written as follows:

$$^{3}m_{t+1} = 2\left(^{3}m_{t}\right) - {}^{3}m_{t-1} + i \tag{10.3.1}$$

By assuming a constant *i* the equation is triangular specific so the perspective designator (superscript 3) can be dropped, thus simplifying equation 10.3.1 to the following form:

$$m_{t+1} = 2m_t - m_{t-1} + i \tag{10.3.2}$$

Equation 10.3.2 can then be used to characterize the various triangularly proportioned triples-lengths by defining an initial magnitude (m_0) for m_{t-1}, a next magnitude (m_1) for m_t, and an increment *i* that results in the observed proportion. To do this, however, the appropriate increment must be determined.

Note that the triangular number series (0, 1, 3, 6, 10, 15, ...) has arithmetically proportioned differences between terms (1, 2, 3, 4, 5, ...) and i=1. If the continuous proportions of r_h and r_d series for near isosceles triples are evaluated in the same way, the increase in the difference between terms is still linear, but the increment is greater than 1: r_h begins 3, 10, 21, 36 and has differences between terms of 7, 11, 15; And r_d begins 6, 15, 28, 45 and has differences between terms of 9, 13, 17. In both cases the difference between terms increases in increments of 4. As is evident, these larger increments result in larger translations within the continuous triangular proportion.

Yet every other triangular number corresponds to every other linear number (i.e. scalar translations of 2). So how do the scalar translations of 2 result in triangular increments that increase by 4? Notice that the m_{t-1} term of the r_h and r_d proportions is in reality the m_{t-2} term of the full triangular number series and that the m_{t+1} term of the r_h and r_d proportions is in reality the m_{t+2} term of the full triangular number series. Per equation 10.3.2, the m_{t-1} term is now always 2 less than it is in the full triangular number series, and the m_{t+1} is always 2 more instead of 1 more. As a result of these two factors i increases from 1 to 4 in these modified triangular proportions. Since all triangular proportions associated with the case of δ_o=1 are linked to the proportions of r_h and r_d they all have an i equal to 4; these include the following proportions given in Figure 46: r_h, r_d, h, d, and δ_e.

Example: $m_{t+1} = 2m_t - m_{t-1} + i$ in terms of r_h becomes $r_{h2} = 2r_{h1} - r_{h0} + i$

Let $r_{h0} = 3$, $r_{h1} = 10$, and $i = 4$

Therefore $r_{h2} = 2 \times 10 - 3 + 4 = 21 \quad \Rightarrow \quad 3, 10, 21, \ldots, r_{h(t-1)}, r_{h(t)}, r_{h(t+1)}$

In the case of δ_e=2: r_v begins 2, 12, 30, 56 and has differences between terms of 10, 18, 26, and r_d begins 6, 20, 42, 72 and has differences between terms of 14, 22, 30. In both cases the differences between terms grows linearly, but increases in increments of 8 (i=8). It follows then that all the triangular proportions of Figure 42 (r_v, r_d, v, d, and δ_o) increment by 8.

Example: $m_{t+1} = 2m_t - m_{t-1} + i$ in terms of r_v becomes $r_{v2} = 2r_{v1} - r_{v0} + i$

Let $r_{v0} = 2$, $r_{v1} = 12$, and $i = 8$

Therefore $r_{v2} = 2 \times 12 - 2 + 8 = 30 \quad \Rightarrow \quad 2, 12, 30, \ldots, r_{v(t-1)}, r_{v(t)}, r_{v(t+1)}$

Many fixed-delta cases beyond δ_o=1 and δ_e=2 can be tested in order to determine how *i* might vary for the triples proportions under consideration, and in all cases of fixed odd delta, *i* is equal to 4, and in all cases of fixed even delta, *i* is equal to 8. Since all triples proportions are described by either *i*=4 or *i*=8, and since these two increments arise from the δ_o=1 and δ_e=2 classes of triples, it is reasonable to characterize the broader classes of fixed odd and even deltas through case studies of these two classes. Therefore, the coming analysis will be limited to these two representative cases of odd and even fixed-delta classes that are both in continuous proportion with the 3-4-5 triangle.

10.4 Continuous Prime Proportions

The magnitude of the Pythagorean triple diagonal is often prime, and it is always equal to the difference between its ex-circle radius and the in-circle radius beginning with the 5 of the 3-4-5 triangle. The triple diagonal is also always equal to the sum of orthogonal ex-radii r_v and r_h. A second series associated with the same triple that is also often prime is formed from the sum of the ex-radius and the in-radius and begins with 7. It is also always equal to the sum of the orthogonal sides *v* and *h*. So these often-prime series are both formed by summing orthogonal linear and triangular continuous proportions. Therefore, I categorize the continuous proportions of triples diagonals as *circle-prime* series (r_v + r_h) and their triangular *twin* continuous proportions as *triangle-prime* series (*v* + *h*). Since these often-prime continuous proportions are formed from orthogonal triangular and linear continuous proportions, I want to know the relative spatial-temporal orientation of these constituent proportions.

The continuous triangular proportions of circle-prime and triangle-prime series are both formed exclusively from a distinct class of numbers composed of primes and products of primes. Within a given often-prime continuous proportion of this type the numbers that are not prime are always products of primes that precede them in the continuous proportion. This is because each series term within these often-prime continuous proportions participates in its own unique component continuous proportion (*sub*-series) in addition to participating in the main continuous proportion of its circle-prime or triangle-prime *carrier*-series. So each carrier-series supports nested sub-series affiliated with each carrier series term. This phenomenon is analogous to how FM radio functions where the carrier-series are like the carrier frequencies you tune to and the sub-series are like modulated frequencies that ride upon the carrier frequencies transmitting content.

The numbers that belong to a given sub-series are all divisible by the carrier-series term that they are in continuous proportion with. These continuous proportions will explain why some carrier-series terms are prime while others are not. They will also help determine the spatial-temporal orientation of their formative triangular and linear proportions.

I refer to the distinct class of numbers that participate in these often-prime continuous proportions as *prime-phase* number. The prime-phase number class is composed of those integers that are either one less or one more than multiples of six. This class of number is divided into a primary chromo-numeric phase consisting of those integers that are greater by one than any multiple of six, and a secondary chromo-numeric phase consisting of those integers that are less by one than any multiple of six, as detailed in Figure 48.* This includes both 1 and -1† since 6x0±1 satisfies the prime phase requisite condition of 6n±1. I differ from convention since I don't consider the numbers 2 and 3 to be *true* primes because they alone among conventional primes do not belong to this class of number.

6n-1	-1	5	11	17	23	29	35	41	47	53	59	65	71	77	83	89	95	101	107	113	119	125	131	137
6n	0	6	12	18	24	30	36	42	48	54	60	66	72	78	84	90	96	102	108	114	120	126	132	138
6n+1	1	7	13	19	25	31	37	43	49	55	61	67	73	79	85	91	97	103	109	115	121	127	133	139
	143	149	155	161	167	173	179	185	191	197	203	209	215	221	227	233	239	245	251	257	263	269	275	281
	144	150	156	162	168	174	180	186	192	198	204	210	216	222	228	234	240	246	252	258	264	270	276	282
	145	151	157	163	169	175	181	187	193	199	205	211	217	223	229	235	241	247	253	259	265	271	277	283
	287	293	299	305	311	317	323	329	335	341	347	353	359	365	371	377	383	389	395	401	407	413	419	425
	288	294	300	306	312	318	324	330	336	342	348	354	360	366	372	378	384	390	396	402	408	414	420	426
	289	295	301	307	313	319	325	331	337	343	349	355	361	367	373	379	385	391	397	403	409	415	421	427
	431	437	443	449	455	461	467	473	479	485	491	497	503	509	515	521	527	533	539	545	551	557	563	569
	432	438	444	450	456	462	468	474	480	486	492	498	504	510	516	522	528	534	540	546	552	558	564	570
	433	439	445	451	457	463	469	475	481	487	493	499	505	511	517	523	529	535	541	547	553	559	565	571

Figure 48. Primary and secondary prime phases: the primary phase consists of all integers that belong to the group 6n+1 and the secondary phase consists of all integers that belong to the group 6n-1, where n may be any integer. Prime numbers are underlined (primes courtesy of primes.utm.edu).

There are those that would argue that all numbers are composed of prime divisors. Yet the numbers 2 and 3 are not part of either primary or secondary prime phases as defined. These numbers can be likened instead to the progenitors of primes, since they are component radii of the 3-4-5 triangle that form its diagonal ex-radius: the *perfect number* 6. They are conventionally considered prime because of their indivisibility by any other number than one, which in their cases can be attributed to their small size. Furthermore, the number two generates every even number, none of which are prime unless the

*The primary and secondary prime-phase relationship to multiples of six gives rise to the mathematical phenomenon know as *twin primes*, the cases where pairs of primes differ by 2.

†Figure 48 shows -1 belonging to the secondary prime phase (violet) since -1+9=8 per the Chapter 5 discussion regarding *absolute polarity*.

number two is counted among them. And the number three belongs neither to the primary nor to the secondary phase, unlike all other primes. So, as there are a great many numbers said to be composed of primes because they are composed of the divisors two and/or three, these numbers are, by the preceding definition, not composed solely of *true* prime divisors, whereas those of the primary and secondary prime phases are.

These prime phases can be distinguished within given circle-prime or triangle-prime continuous proportion as shown in Figure 49, wherein the unit (1) is both added and subtracted from each series term and the resulting quantity is divided by six. When one less than the term divided by six is non-fractional, the term belongs to the primary prime phase. And when one more than the term divided by six is non-fractional, the term belongs to the secondary prime phase. Figure 49 is provided to illustrate the continuous proportions of circle and triangle primes for the cases of δ_o=1 and δ_e=2. Notice that for the case of δ_o=1 two thirds of numeric species belong to the primary prime phase, and one third to the secondary. In the case of δ_e=2, the opposite is true.

For all circle and triangle prime continuous proportions analyzed in this way the alternation between primary and secondary primes phases is regular and all have similar 9-fold analog chromo-numeric transformations. These analog transformations all consist of three step numeric polarity inversions between primary and secondary prime phases. In every case there are similar patterns of mirroring polarity sequences. These numeric patterns are interesting, but they don't provide any direct hints about the nested sub-series that originate from each series term that generate the non-prime terms within the prime phase proportion of the carrier-series. So there must still be some deeper underlying prime symmetry not exposed so far in the analysis.

To highlight these nested sub-series the first step is to analyze the divisors of the carrier-series as shown in Figures 50a (δ_o=1) and 50b (δ_e=2). Consider the first two often-prime series associated with δ_o=1: the circle-prime series commencing with 5 whose terms are composed of r_v and r_h, and the triangle-prime series commencing with 7 whose terms are composed of *v* and *h*. A given carrier series term that is not prime is always divisible by a preceding carries series term with which it is in continuous proportion within a separate nested sub-series. This observation is most easily demonstrated by dividing each series term by the same preceding one. So for example to find those terms that are divisible by the prime length of 5, divide all terms by 5. Next divide all terms by the second length, (13), then divide by the third, (25), and so on. Dividing by the sequence of series terms shows the prime divisors that compose each non-prime series term. I've demonstrated the numeric analysis process with the first three terms of each series since they are sufficient to characterize the continuous proportions of thereof.

$\delta_o = 1$							
prime		$(p+1)/6$	$(p-1)/6$	prime		$(p+1)/6$	$(p-1)/6$
5	5	**1**	2/3	7	7	1 1/3	**1**
13	4	2 1/3	**2**	17	8	**3**	2 2/3
25	7	4 1/3	**4**	31	4	5 1/3	**5**
41	5	**7**	6 2/3	49	4	8 1/3	**8**
61	7	10 1/3	**10**	71	8	**12**	11 2/3
85	4	14 1/3	**14**	97	7	16 1/3	**16**
113	5	**19**	18 2/3	127	1	21 1/3	**21**
145	1	24 1/3	**24**	161	8	**27**	26 2/3
181	1	30 1/3	**30**	199	1	33 1/3	**33**
221	5	**37**	36 2/3	241	7	40 1/3	**40**
265	4	44 1/3	**44**	287	8	**48**	47 2/3
313	7	52 1/3	**52**	337	4	56 1/3	**56**
365	5	**61**	60 2/3	391	4	65 1/3	**65**
421	7	70 1/3	**70**	449	8	**75**	74 2/3
481	4	80 1/3	**80**	511	7	85 1/3	**85**
545	5	**91**	90 2/3	577	1	96 1/3	**96**
613	1	102 1/3	**102**	647	8	**108**	107 2/3
685	1	114 1/3	**114**	721	1	120 1/3	**120**
761	5	**127**	126 2/3	799	7	133 1/3	**133**
841	4	140 1/3	**140**	881	8	**147**	146 2/3
925	7	154 1/3	**154**	967	4	161 1/3	**161**
1013	5	**169**	168 2/3	1057	4	176 1/3	**176**
1105	7	184 1/3	**184**	1151	8	**192**	191 2/3
1201	4	200 1/3	**200**	1249	7	208 1/3	**208**
1301	5	**217**	216 2/3	1351	1	225 1/3	**225**
1405	1	234 1/3	**234**	1457	8	**243**	242 2/3
1513	1	252 1/3	**252**	1567	1	261 1/3	**261**
1625	5	**271**	270 2/3	1681	7	280 1/3	**280**
1741	4	290 1/3	**290**	1799	8	**300**	299 2/3
1861	7	310 1/3	**310**	1921	4	320 1/3	**320**
1985	5	**331**	330 2/3	2047	4	341 1/3	**341**
2113	7	352 1/3	**352**	2177	8	**363**	362 2/3
2245	4	374 1/3	**374**	2311	7	385 1/3	**385**
2381	5	**397**	396 2/3	2449	1	408 1/3	**408**
2521	1	420 1/3	**420**	2591	8	**432**	431 2/3
2665	1	444 1/3	**444**	2737	1	456 1/3	**456**
2813	5	**469**	468 2/3	2887	7	481 1/3	**481**
2965	4	494 1/3	**494**	3041	8	**507**	506 2/3
3121	7	520 1/3	**520**	3199	4	533 1/3	**533**
3281	5	**547**	546 2/3	3361	4	560 1/3	**560**
3445	7	574 1/3	**574**	3527	8	**588**	587 2/3
3613	4	602 1/3	**602**	3697	7	616 1/3	**616**

$\delta_e = 2$							
prime		$(p+1)/6$	$(p-1)/6$	prime		$(p+1)/6$	$(p-1)/6$
5	5	**1**	2/3	7	7	1 1/3	**1**
17	8	**3**	2 2/3	23	5	**4**	3 2/3
37	1	6 1/3	**6**	47	2	**8**	7 2/3
65	2	**11**	10 2/3	79	7	13 1/3	**13**
101	2	**17**	16 2/3	119	2	**20**	19 2/3
145	1	24 1/3	**24**	167	5	**28**	27 2/3
197	8	**33**	32 2/3	223	7	37 1/3	**37**
257	5	**43**	42 2/3	287	8	**48**	47 2/3
325	1	54 1/3	**54**	359	8	**60**	59 2/3
401	5	**67**	66 2/3	439	7	73 1/3	**73**
485	8	**81**	80 2/3	527	5	**88**	87 2/3
577	1	96 1/3	**96**	623	2	**104**	103 2/3
677	2	**113**	112 2/3	727	7	121 1/3	**121**
785	2	**131**	130 2/3	839	2	**140**	139 2/3
901	1	150 1/3	**150**	959	5	**160**	159 2/3
1025	8	**171**	170 2/3	1087	7	181 1/3	**181**
1157	5	**193**	192 2/3	1223	8	**204**	203 2/3
1297	1	216 1/3	**216**	1367	8	**228**	227 2/3
1445	5	**241**	240 2/3	1519	7	253 1/3	**253**
1601	8	**267**	266 2/3	1679	5	**280**	279 2/3
1765	1	294 1/3	**294**	1847	2	**308**	307 2/3
1937	2	**323**	322 2/3	2023	7	337 1/3	**337**
2117	2	**353**	352 2/3	2207	2	**368**	367 2/3
2305	1	384 1/3	**384**	2399	5	**400**	399 2/3
2501	8	**417**	416 2/3	2599	7	433 1/3	**433**
2705	5	**451**	450 2/3	2807	8	**468**	467 2/3
2917	1	486 1/3	**486**	3023	8	**504**	503 2/3
3137	5	**523**	522 2/3	3247	7	541 1/3	**541**
3365	8	**561**	560 2/3	3479	5	**580**	579 2/3
3601	1	600 1/3	**600**	3719	2	**620**	619 2/3
3845	2	**641**	640 2/3	3967	7	661 1/3	**661**
4097	2	**683**	682 2/3	4223	2	**704**	703 2/3
4357	1	726 1/3	**726**	4487	5	**748**	747 2/3
4625	8	**771**	770 2/3	4759	7	793 1/3	**793**
4901	5	**817**	816 2/3	5039	8	**840**	839 2/3
5185	1	864 1/3	**864**	5327	8	**888**	887 2/3
5477	5	**913**	912 2/3	5623	7	937 1/3	**937**
5777	8	**963**	962 2/3	5927	5	**988**	987 2/3
6085	1	1014 1/3	**1014**	6239	2	**1040**	1039 2/3
6401	2	**1067**	1066 2/3	6559	7	1093 1/3	**1093**
6725	2	**1121**	1120 2/3	6887	2	**1148**	1147 2/3
7057	1	1176 1/3	**1176**	7223	5	**1204**	1203 2/3

Figure 49. Numeric analysis of *circle* ($r_v + r_h$) and *triangle* ($v + h$) continuous prime-phase proportions for cases of δ_o=1 and δ_e=2: For δ_o=1 two thirds of numeric species belong to the primary prime phase and one third belong to the secondary prime phase. For δ_e=2 one third of numeric species belong to the primary prime phase and two thirds belong to the secondary prime phase. In both cases 9-fold analog chromo-numeric transformations can be seen. Primes are underlined.

The frequencies at which terms arise that are divisible by the same original series term are easy to spot. These succeeding non-prime terms appear at regular intervals that correspond to the magnitudes of the prime-length's constituents (r_v and r_h or *v* and *h*). For example: the first length of the circle prime series (5) is composed of the magnitudes r_v (2) and r_h (3), and terms divisible by 5 are found to alternate 2 and 3 terms apart. So when every other term divisible by 5 is assigned to separate r_v and r_h continuous proportions (sub-series), the frequency of terms divisible by 5 becomes 1/5th. The separation (*offset*) between the terms of these two sub-series matches the component magnitudes for the term-1 sub-series pair.

Beyond the first term there continues to be two interwoven proportions with divisible term frequencies that are the inverse of term length, but their original constituent lengths (r_v and r_h or *v* and *h*) no longer correspond to the offsets between them. Rather, the offset between the first and second interwoven sub-series always increases with triangular proportion (*triangular offset*). The converse offset between the second and first sub-series increases with arithmetic proportion (*linear offset*). The linear offset always increases by 2 relative to the preceding offset, so it can be used to calculate triangular offsets for sub-series beyond the term-1 divisor, since the linear offset and the triangular offset always sum to the original divisor term (*p*). These offsets are shown in Figure 50a.

Distinctions between these sub-series hint at the symmetry and orientation of the carrier series' formative triangular and linear proportions: In the case of δ_e=2 shown in Figure 50b, the size of the initial triangular and linear offsets are reversed relative to the δ_o=1 case. These differences suggest that the continuous proportions of δ_o=1 and δ_e=2 cases are oppositely oriented consistent with the difference between triangle and circle equations for odd and even delta highlighted previously. In the case of δ_o=1 r_v is linear and r_h is triangular, whereas for δ_e=2 r_v is triangular and r_h is linear. What remains to be determined is whether there is a spatial-temporal relationship between this orthogonal triangular-linear dimensional symmetry.

As I've shown before, the means I use to characterize a given continuous proportion is to perform a difference analysis to determine how succeeding terms increase in magnitude. Take for instance Figure 50a, the circle prime term-1 divisor sub-series: 5, 29, 73, 137, 221, ... whose term differences are 24, 44, 64, 84, ... In this example it is clear that the difference between terms is incremented by 20, which is the increment (*i*) introduced in equation 10.3.2 multiplied by the prime divisor (*p*): $ip = 4 \times 5$. So the continuous proportions of these divisor sub-series can be described by modifying equation 10.3.2 to account for this relationship as follows:

$$m_{t+1} = 2m_t - m_{t-1} + ip \qquad (10.4.1)$$

$\delta_o = 1$ (i=4: series 1, series 2)							
term (p):	5	13	25	term (p):	7	17	31
triang. offset:	2	8	18	triang. offset:	3	11	23
linear offset:	3	5	7	linear offset:	4	6	8
prime	÷ term 1	÷ term 2	÷ term 3	prime	÷ term 1	÷ term 2	÷ term 3
5	1	5/13	1/5	7	1	7/17	7/31
13	2 3/5	1	13/25	17	2 3/7	1	17/31
25	5	1 12/13	1	31	4 3/7	1 14/17	1
41	8 1/5	3 2/13	1 16/25	49	7	2 15/17	1 18/31
61	12 1/5	4 9/13	2 11/25	71	10 1/7	4 3/17	2 9/31
85	17	6 7/13	3 2/5	97	13 6/7	5 12/17	3 4/31
113	22 3/5	8 9/13	4 13/25	127	18 1/7	7 8/17	4 3/31
145	29	11 2/13	5 4/5	161	23	9 8/17	5 6/31
181	36 1/5	13 12/13	7 6/25	199	28 3/7	11 12/17	6 13/31
221	44 1/5	17	8 21/25	241	34 3/7	14 3/17	7 24/31
265	53	20 5/13	10 3/5	287	41	16 15/17	9 8/31
313	62 3/5	24 1/13	12 13/25	337	48 1/7	19 14/17	10 27/31
365	73	28 1/13	14 3/5	391	55 6/7	23	12 19/31
421	84 1/5	32 5/13	16 21/25	449	64 1/7	26 7/17	14 15/31
481	96 1/5	37	19 6/25	511	73	30 1/17	16 15/31
545	109	41 12/13	21 4/5	577	82 3/7	33 16/17	18 19/31
613	122 3/5	47 2/13	24 13/25	647	92 3/7	38 1/17	20 27/31
685	137	52 9/13	27 2/5	721	103	42 7/17	23 8/31
761	152 1/5	58 7/13	30 11/25	799	114 1/7	47	25 24/31
841	168 1/5	64 9/13	33 16/25	881	125 6/7	51 14/17	28 13/31
925	185	71 2/13	37	967	138 1/7	56 15/17	31 6/31
1013	202 3/5	77 12/13	40 13/25	1057	151	62 3/17	34 3/31
1105	221	85	44 1/5	1151	164 3/7	67 12/17	37 4/31
1201	240 1/5	92 5/13	48 1/25	1249	178 3/7	73 8/17	40 9/31
1301	260 1/5	100 1/13	52 1/25	1351	193	79 8/17	43 18/31
1405	281	108 1/13	56 1/5	1457	208 1/7	85 12/17	47
1513	302 3/5	116 5/13	60 13/25	1567	223 6/7	92 3/17	50 17/31
1625	325	125	65	1681	240 1/7	98 15/17	54 7/31
1741	348 1/5	133 12/13	69 16/25	1799	257	105 14/17	58 1/31
1861	372 1/5	143 2/13	74 11/25	1921	274 3/7	113	61 30/31
1985	397	152 9/13	79 2/5	2047	292 3/7	120 7/17	66 1/31
2113	422 3/5	162 7/13	84 13/25	2177	311	128 1/17	70 7/31
2245	449	172 9/13	89 4/5	2311	330 1/7	135 16/17	74 17/31
2381	476 1/5	183 2/13	95 6/25	2449	349 6/7	144 1/17	79
2521	504 1/5	193 12/13	100 21/25	2591	370 1/7	152 7/17	83 18/31
2665	533	205	106 3/5	2737	391	161	88 9/31
2813	562 3/5	216 5/13	112 13/25	2887	412 3/7	169 14/17	93 4/31
2965	593	228 1/13	118 3/5	3041	434 3/7	178 15/17	98 3/31
3121	624 1/5	240 1/13	124 21/25	3199	457	188 3/17	103 6/31
3281	656 1/5	252 5/13	131 6/25	3361	480 1/7	197 12/17	108 13/31
3445	689	265	137 4/5	3527	503 6/7	207 8/17	113 24/31
3613	722 3/5	277 12/13	144 13/25	3697	528 1/7	217 8/17	119 8/31

triangular offset — linear offset

Figure 50a. δ_o=1 circle prime and triangle prime continuous proportion divisor analysis: Terms divisible by a given preceding term are separated into two series and are color-coded. The terms of a given separated series occur at a frequency equal to the reciprocal of the divisor term, for instance every 5th term is divisible by 5. The two series associated with the given term are separated by triangular and linear offsets as shown.

δ_e = 2 (*i*=8: series 1, series 2)							
term (*p*):	5	17	37	term (*p*):	7	23	47
triang. offset:	2	13	31	triang. offset:	3	18	40
linear offset:	3	4	6	linear offset:	4	5	7
prime	÷ term 1	÷ term 2	÷ term 3	prime	÷ term 1	÷ term 2	÷ term 3
5	1	5/17	5/37	7	1	7/23	7/47
17	3 2/5	1	17/37	23	3 2/7	1	23/47
37	7 2/5	2 3/17	1	47	6 5/7	2 1/23	1
65	13	3 14/17	1 28/37	79	11 2/7	3 10/23	1 32/47
101	20 1/5	5 16/17	2 27/37	119	17	5 4/23	2 25/47
145	29	8 9/17	3 34/37	167	23 6/7	7 6/23	3 26/47
197	39 2/5	11 10/17	5 12/37	223	31 6/7	9 16/23	4 35/47
257	51 2/5	15 2/17	6 35/37	287	41	12 11/23	6 5/47
325	65	19 2/17	8 29/37	359	51 2/7	15 14/23	7 30/47
401	80 1/5	23 10/17	10 31/37	439	62 5/7	19 2/23	9 16/47
485	97	28 9/17	13 4/37	527	75 2/7	22 21/23	11 10/47
577	115 2/5	33 16/17	15 22/37	623	89	27 2/23	13 12/47
677	135 2/5	39 14/17	18 11/37	727	103 6/7	31 14/23	15 22/47
785	157	46 3/17	21 8/37	839	119 6/7	36 11/23	17 40/47
901	180 1/5	53	24 13/37	959	137	41 16/23	20 19/47
1025	205	60 5/17	27 26/37	1087	155 2/7	47 6/23	23 6/47
1157	231 2/5	68 1/17	31 10/37	1223	174 5/7	53 4/23	26 1/47
1297	259 2/5	76 5/17	35 2/37	1367	195 2/7	59 10/23	29 4/47
1445	289	85	39 2/37	1519	217	66 1/23	32 15/47
1601	320 1/5	94 3/17	43 10/37	1679	239 6/7	73	35 34/47
1765	353	103 14/17	47 26/37	1847	263 6/7	80 7/23	39 14/47
1937	387 2/5	113 16/17	52 13/37	2023	289	87 22/23	43 2/47
2117	423 2/5	124 9/17	57 8/37	2207	315 2/7	95 22/23	46 45/47
2305	461	135 10/17	62 11/37	2399	342 5/7	104 7/23	51 2/47
2501	500 1/5	147 2/17	67 22/37	2599	371 2/7	113	55 14/47
2705	541	159 2/17	73 4/37	2807	401	122 1/23	59 34/47
2917	583 2/5	171 10/17	78 31/37	3023	431 6/7	131 10/23	64 15/47
3137	627 2/5	184 9/17	84 29/37	3247	463 6/7	141 4/23	69 4/47
3365	673	197 16/17	90 35/37	3479	497	151 6/23	74 1/47
3601	720 1/5	211 14/17	97 12/37	3719	531 2/7	161 16/23	79 6/47
3845	769	226 3/17	103 34/37	3967	566 5/7	172 11/23	84 19/47
4097	819 2/5	241	110 27/37	4223	603 2/7	183 14/23	89 40/47
4357	871 2/5	256 5/17	117 28/37	4487	641	195 2/23	95 22/47
4625	925	272 1/17	125	4759	679 6/7	206 21/23	101 12/47
4901	980 1/5	288 5/17	132 17/37	5039	719 6/7	219 2/23	107 10/47
5185	1037	305	140 5/37	5327	761	231 14/23	113 16/47
5477	1095 2/5	322 3/17	148 1/37	5623	803 2/7	244 11/23	119 30/47
5777	1155 2/5	339 14/17	156 5/37	5927	846 5/7	257 16/23	126 5/47
6085	1217	357 16/17	164 17/37	6239	891 2/7	271 6/23	132 35/47
6401	1280 1/5	376 9/17	173	6559	937	285 4/23	139 26/47
6725	1345	395 10/17	181 28/37	6887	983 6/7	299 10/23	146 25/47
7057	1411 2/5	415 2/17	190 27/37	7223	1031 6/7	314 1/23	153 32/47
7397	1479 2/5	435 2/17	199 34/37	7567	1081	329	161

Figure 50b. δ_e=2 circle prime and triangle prime continuous proportion divisor analysis: Terms divisible by a given preceding term are separated into two series and are color-coded. The terms of a given separated series occur at a frequency equal to the reciprocal of the divisor term. The two series associated with the given term are separated by triangular and linear offsets that are oriented opposite to those of the δ_o=1 case.

This new form of continuous proportion is easily verified by applying it to the cases presently under consideration. In other words, let *i* equal what it was for the carrier series, and *p* then is equal to the carrier series term under evaluation. For series 1, let the initial magnitude (m_0) and next magnitude (m_1) equal the first and third terms of the divisor series (since divisible terms alternate between the two paired sub-series), and conversely for series 2 let the initial magnitude (m_0) and next magnitude (m_1) equal the second and fourth terms of the divisor series. The resulting continuous proportions correspond to those detailed in Figure 50a when the two constituent series are interleaved:

CIRCLE PRIME

divisor series 1: $i = 4,\ p = 5, m_0 = 1,\ m_1 = 17 \ \Rightarrow\ 1,\ 17,\ 53,\ 109,\ 185,\ 281,\ 397,\ \ldots$
divisor series 2: $i = 4,\ p = 5, m_0 = 5,\ m_1 = 29 \ \Rightarrow\ 5,\ 29,\ 73,\ 137,\ 221,\ 325,\ 449,\ \ldots$

TRIANGLE PRIME

divisor series 1: $i = 4,\ p = 7, m_0 = 1,\ m_1 = 23 \ \Rightarrow\ 1,\ 23,\ 73,\ 151,\ 257,\ 391,\ 553,\ \ldots$
divisor series 2: $i = 4,\ p = 7, m_0 = 7,\ m_1 = 41 \ \Rightarrow\ 7,\ 41,\ 103,\ 193,\ 311,\ 457,\ 631,\ \ldots$

The preceding findings suggest that these sub-series pairs emanate from series terms, and propagate within their carrier-series as distinct continuous proportions. But what are these continuous proportions? If the divisor proportion of a given circle or triangle prime sub-series is incremented by *ip*, then it stands to reason that the sub-series itself would be incremented by ip^2 since the divisor proportion has been divided by *p*. Then, the following equation describes the continuous triangular proportions of the nested pairs of sub-series within each carrier-series:

$$m_{t+1} = 2m_t - m_{t-1} + ip^2 \qquad (10.4.2)$$

To better understand the geometric relationship between these paired continuous proportions I next analyze their symmetries using modulus-9 and term difference analysis. Figure 51 shows this analysis for the paired sub-series (series 1, series 2) associated with the first term of both circle and triangle prime carrier-series for both δ_o=1 and δ_e=2. The figure therefore includes the four series associated with the first primary phase prime (7) and the four series associated with the first secondary phase prime (5). The analog polarity patterns of these series are of the same general form as those of the carrier series shown in Figure 49.

$\delta_o = 1$								$\delta_e = 2$							
prime: ip^2=100				prime: ip^2=196				prime: ip^2=200				prime: ip^2=392			
series 1		series 2		series 1		series 2		series 1		series 2		series 1		series 2	
5	5	25	7	7	7	49	4	5	5	65	2	7	7	119	2
85	4	145	1	161	8	287	8	145	1	325	1	287	8	623	2
265	4	365	5	511	7	721	1	485	8	785	2	959	5	1519	7
545	5	685	1	1057	4	1351	1	1025	8	1445	5	2023	7	2807	8
925	7	1105	7	1799	8	2177	8	1765	1	2305	1	3479	5	4487	5
1405	1	1625	5	2737	1	3199	4	2705	5	3365	8	5327	8	6559	7
1985	5	2245	4	3871	1	4417	7	3845	2	4625	8	7567	7	9023	5
2665	1	2965	4	5201	8	5831	8	5185	1	6085	1	10199	2	11879	8
3445	7	3785	5	6727	4	7441	7	6725	2	7745	5	13223	2	15127	7
4325	5	4705	7	8449	7	9247	4	8465	5	9605	2	16639	7	18767	2
5305	4	5725	1	10367	8	11249	8	10405	1	11665	1	20447	8	22799	2
6385	4	6845	5	12481	7	13447	1	12545	8	13925	2	24647	5	27223	7
7565	5	8065	1	14791	4	15841	1	14885	8	16385	5	29239	7	32039	8
8845	7	9385	7	17297	8	18431	8	17425	1	19045	1	34223	5	37247	5
10225	1	10805	5	19999	1	21217	4	20165	5	21905	8	39599	8	42847	7
continued proportions of differences between terms per ip^2															
-320		*-280*		*-630*		*-546*		*-660*		*-540*		*-1288*		*-1064*	
-220		*-180*		*-434*		*-350*		*-460*		*-340*		*-896*		*-672*	
-120		*-80*		*-238*		*-154*		*-260*		*-140*		*-504*		*-280*	
-20		*20*		*-42*		*42*		*-60*		*60*		*-112*		*112*	
80		120		154		238		140		260		280		504	
180		220		350		434		340		460		672		896	
280		320		546		630		540		660		1064		1288	

Figure 51. Isolated sub-series continuous proportion of the circle and triangle prime series terms: Each sub-series is affiliated with their respective prime divisors (5 and 7) through the ip^2 continuous triangular proportion. Modulus-9 symmetry and the continued proportion of differences between terms indicate that series-1/series-2 pairs are opposite directional components of the same continuous proportion.

Also, the polarity patterns of each pair of sub-series are actually the same, although they appear to be reverse. So, if a given series-2 is flipped and concatenated to series-1, they appear continuous. And, if one extends the continuous proportions of the differences between terms of a given series-1 and series-2 pair using the ip^2 quantity, they appear to be the same except their continuous proportions are oriented in opposite direction to one another (see Figure 51, lower section). These findings show that the series-1 series-2 pairs are actually objective and subjective components of the same continuous proportion, spatial and temporal components perhaps. As affirmation of the Figure 51 findings, Figure 52 combines ip^2 sub-series pairs detailed in Figure 51 into one continuous proportion with positive (series-1) and negative (series-2) components emanating from the original carrier series term. It also includes the second term ip^2 sub-series shown in Figures 50a and 50b.

$\delta_o = 1$								$\delta_e = 2$							
prime				prime				prime				prime			
ip^2=100		ip^2=676		ip^2=196		ip^2=1156		ip^2=200		ip^2=2312		ip^2=392		ip^2=4232	
3785	5	26221	4	7441	7	44999	8	7745	5	91205	8	15127	7	167279	5
2965	4	20605	4	5831	8	35377	7	6085	1	71825	5	11879	8	131767	7
2245	4	15665	5	4417	7	26911	1	4625	8	54757	1	9023	5	100487	2
1625	5	11401	7	3199	4	19601	8	3365	8	40001	5	6559	7	73439	8
1105	7	7813	1	2177	8	13447	1	2305	1	27557	8	4487	5	50623	7
685	1	4901	5	1351	1	8449	7	1445	5	17425	1	2807	8	32039	8
365	5	2665	1	721	1	4607	8	785	2	9605	2	1519	7	17687	2
145	1	1105	7	287	8	1921	4	325	1	4097	2	623	2	7567	7
25	7	221	5	49	4	391	4	65	2	901	1	119	2	1679	5
5	5	13	4	7	7	17	8	5	5	17	8	7	7	23	5
85	4	481	4	161	8	799	7	145	1	1445	5	287	8	2599	7
265	4	1625	5	511	7	2737	1	485	8	5185	1	959	5	9407	2
545	5	3445	7	1057	4	5831	8	1025	8	11237	5	2023	7	20447	8
925	7	5941	1	1799	8	10081	1	1765	1	19601	8	3479	5	35719	7
1405	1	9113	5	2737	1	15487	7	2705	5	30277	1	5327	8	55223	8
1985	5	12961	1	3871	1	22049	8	3845	2	43265	2	7567	7	78959	2
2665	1	17485	7	5201	8	29767	4	5185	1	58565	2	10199	2	106927	7
3445	7	22685	5	6727	4	38641	4	6725	2	76177	1	13223	2	139127	5
4325	5	28561	4	8449	7	48671	8	8465	5	96101	8	16639	7	175559	5

Figure 52. Merged ip^2 continuous proportion of first two terms for both circle and triangle prime series for fixed δ_o=1 and δ_e=2.

Owing to the complexity of these proportions considerable computational resources would be needed to fully characterize their nuances. Notice for instance that at regular intervals the proportion of the second term contains terms from the proportion of the first term and that positive and negative terms are systematically interchanged between the two. This indicates that, although isolated, the symmetries of their respective proportions are intricately interwoven, which is to be expected since they are nested components of the same carrier proportion. Also notice that, in addition to every term being divisible by the original series term p, at a frequency of p terms of a given sub-series are divisible by p^2; and at a frequency of p^2 terms are divisible by p^3 and so on. This can be seen by repeatedly dividing the results of division by p. These attributes illustrate the richness of the symmetry under consideration. Yet, however intriguing these detailed observations are, they are tangential to the present line of reasoning since, in the main, the derivations of the *i*, *ip*, and ip^2 continuous triangular proportions have solved the mystery of *continuous prime proportions* including why some prime phase numbers are prime while others are not.

What remains unconsidered is the bi-directionality of the ip^2 continuous proportion, its sources and physical analogies. Of particular interest is how the two components of the ip^2 proportions relate to their formative constituent symmetries. Recall that circle and triangle prime series are composed of orthogonal constituent proportions that originate from the linear and triangular dimensional perspectives of Pascal's Triangle; these perspectives can be affiliated with the relative orientation of the two components of the ip^2 continuous proportion through the offsets of the two components. This means that if one orthogonal constituent perspective is spatial the other is temporal.* So for instance, if a given triples triangular proportion is spatial, its orthogonal linear proportion must be temporal.

It's reasonable to propose, then, that if series-1 is spatially oriented for the δ_o class it is temporally oriented for the δ_e class, and vice versa for series-2, since the two classes are oppositely oriented in the same way as are the two sub-series pairs. As affirmation of this interpretation, compare the ip^2 continuous proportions for the first terms (5 and 7) of the δ_o=1 series with those of δ_e=2. In the case of δ_o=1 series-2s begin 25, **145** and 49, **287** and in the case of δ_e=2 series-1s begin 5, **145** and 7, **287**. The second terms are identical in both cases, which suggests that the orientations of δ_e=2 sub-series are reverse from δ_o=1 sub-series. This is because their proportions skip over the first terms of series-2s such that 5, 25, 145 (δ_o=1) becomes 5, 145 (δ_e=2) and 7, 49, 287 (δ_o=1) becomes 7, 287 (δ_e=2) when ip^2 is doubled.† This observation affirms that the continuous proportions of the δ_e class are oppositely oriented to those of the δ_o class. It suggests that for the δ_o=1 case the formative triangular proportion is spatially oriented and the formative linear proportion is temporally oriented, and that the opposite is true for the δ_e=2 case. But these observations only provide a second-hand inference in the sense that they are symmetric effects. To understand the underlying cause the geometry of the governing 3-4-5 triangle must be interpreted within the spatial-temporal dimensional context of Pascal's Triangle.

---

*One might be tempted to discount this affiliation because the linear and triangular offsets of the two component proportions deviate from their component lengths beyond the first term (see Figures 50a and 50b). Recall, however, that each subsequent term of the circle and triangle primes series corresponding to either δ_o=1 and δ_e=2 is also the first term of its own continuous proportion for higher deltas of each class. And in every case a given first term's complementary constituent lengths can be affiliated with their ip^2 objective and subjective component proportions by their offsets.

†Recall that i = 4 for δ_o = 1 and i = 8 for δ_e = 2 so for equal p the ip^2 for δ_e = 2 is double the ip^2 for δ_o = 1.

10.5 PRIMORDIAL FIELD GEOMETRY

As the preceding analysis has demonstrated, the continuous proportions the ex-circles emanating from the singular (r=1) in-circle of the 3-4-5 triangle correspond to the continuous arithmetic and triangular proportions of Pascal's Triangle. The governing complementary cuboctahedral rotational perspective of Pascal's Triangle suggests that the triangular ex-circle proportion is spatial and the linear ex-circle proportion is temporal since the third super-symmetric spatial/temporal field interval harbors the constrained singular origins of both these dimensions, as shown in Figure 53. The dimensional symmetry of this interval therefore conforms to the orthogonal triangular/linear (spatial/temporal) symmetry of the in-circle of the 3-4-5 triangle.

Recall from Chapter 3 that dimensional constraints are cumulative. Hence, the in-circle must be constrained to inversion between the continuously equal singular spatial and temporal proportions and the linear spatial proportion in addition to the triangular spatial and linear temporal proportions of its ex-circles. Assuming this geometry harbors all the symmetry elements of the third super-symmetric field interval, it is reasonable to expect that the symmetry of the in-circle also has the capacity to adopt the anti-field perspective like the symmetry of Pascal's Triangle.

Owing to the 2:1 (spatial:temporal) constraint asymmetry of the governing continuous golden proportion, this same interval marks the completion of the primordial order field analog from this integral rotational perspective. Additionally, it is the interval where the origins of spatial-triangular and temporal-linear translational symmetries superposition. So it seems logical to entertain the possibility that the dimensional symmetry that governs electromagnetism could evolve from this spatial/temporal nexus.

The in-circle of the 3-4-5 triangle together with its diagonal ex-circle (the perfect number 6) manifest this dual spatial and temporal symmetry within the prime phase field they govern (6n±1). So it naturally follows that the bi-directional continuous ip^2 proportions of prime-phase diagonal species should also harbor the underlying spatial/temporal dimensional symmetry of the in-circle of the 3-4-5 triangle they originate from. Notice that, like the singular origin, the prime-phase species at the origin of continuous ip^2 proportions are centers of inversion. And since they are asymmetrically constrained like the singular origin of the continuous golden proportion, these centers of inversion reside on either spatial or temporal sides of the true inversion center (0) – this is reflected by the asymmetry of continuous proportions shown in Figure 52.

Since δ_o=1 continuous proportions emanating from the in-circle of the 3-4-5 triangle correlate directly to the governing spatial/temporal (triangular/linear) dimensional symmetry of Pascal's Triangle, it seems reasonable to assume that all fixed δ_o proportions are triangular-spatial and linear-temporal. And, it follows that the asymmetric centers (*p*) of the fixed δ_o bi-directional continuous ip^2 proportions are on the spatial side of the inversion center since they are at the beginning of the half of the proportion distinguished from the other half by its *triangular off-set*.

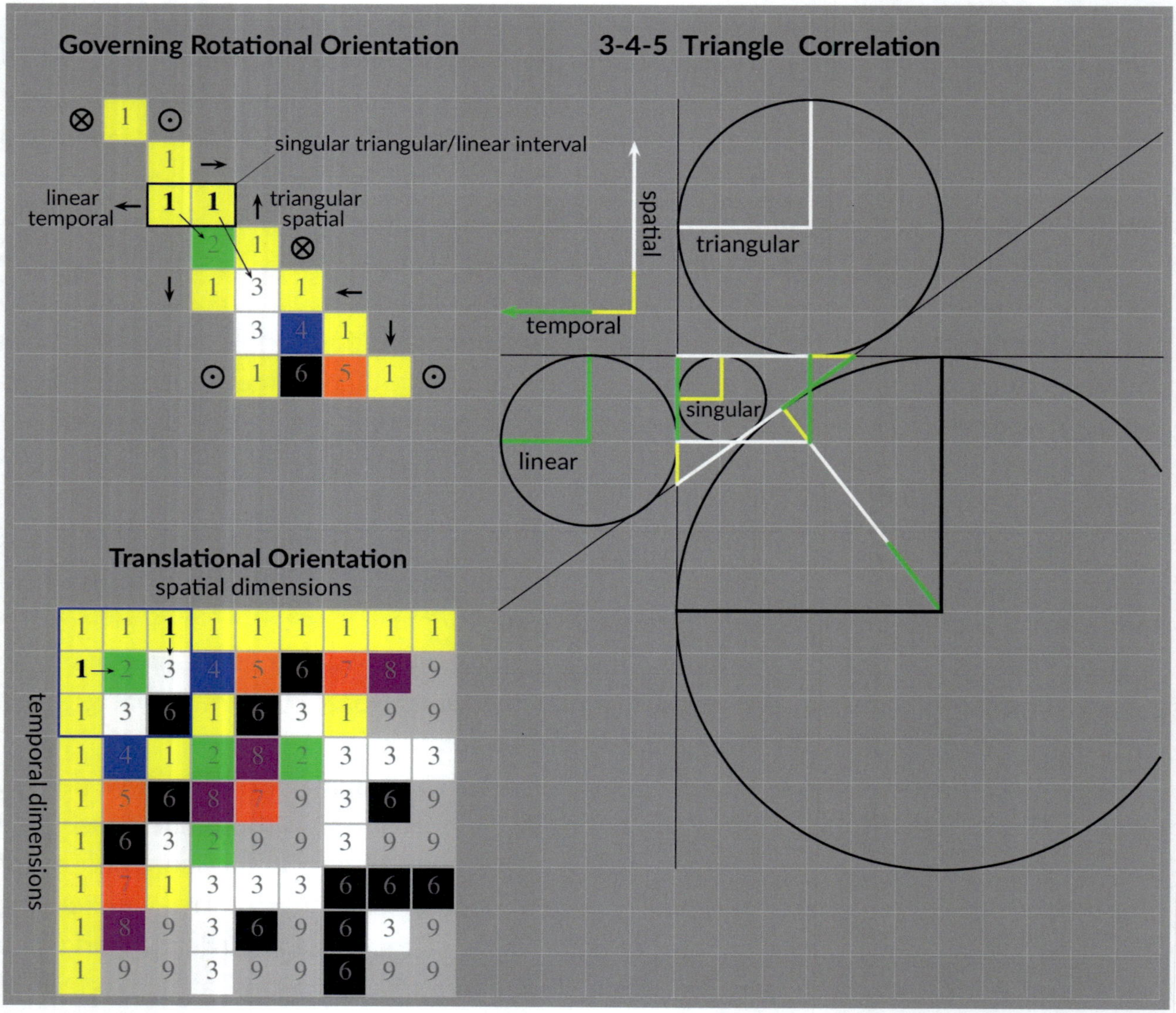

Figure 53. Singular triangular/linear interval perspective: Upper-left the 2:1 (spatial:temporal) constraint asymmetry aligns the triangular-spatial and linear-temporal perspectives of the point to the same spatial/temporal interval. Lower-left this singular triangular/linear interval is shown as the culminating state of the primordial-order field analog. To the right the orthogonal triangular (spatial) and linear (temporal) perspectives of the 3-4-5 triangle combine to form the spatial/temporal proportions the diagonal.

The continuous proportions of the triples of the δ_e class are oriented opposite those of δ_o, suggesting that the triples of the δ_e class are temporal/spatial (assuming that the spatial and temporal orientation of the 3-4-5 triangle is the same for both δ_o=1 and δ_e=2 since both are perspectives of the same triangle in-circle). It follows then, that the asymmetric centers (p) of the fixed δ_e bi-directional continuous ip^2 proportions are on the temporal side of the neutral inversion center (0). Therefore, as the origins of both δ_o and δ_e bi-directional continuous ip^2 proportions, it follows these prime-phase centers harbor both spatial/temporal and temporal/spatial symmetries.

Notice that for the case of δ_o=1 every linear temporal translation of 1 is accompanied by 2 triangular spatial. Assuming the same constraint asymmetry applies in the case of δ_e=2 seems reasonable since both geometries are in continuous proportion with the same 3-4-5 triangle orientation. And as both forms of continuous proportion arise from the same triangle, it follows that for both δ_o=1 and δ_e=2 perspectives r_v remains temporal and r_h spatial and that for the δ_e=2 perspective r_v is triangular and r_h linear (reverse of δ_o=1). Additionally, for δ_e=2 the spatial perspective is linear with arithmetic translations of 2, and the temporal perspective is triangular with translations corresponding to double the case of δ_o=1.[*] This suggests that the perspective maintains the same triangular:linear asymmetry since double triangular translations correspond with double linear translations. Halving the δ_e=2 triangular proportions highlights that the interleaving triangular planes are reversed between the δ_o=1 and δ_e=2 cases. And since triangular orientations are reverse between spatial and temporal perspectives, the reversal of continuous triangular proportions conforms to the reversal of spatial-temporal orientation.[†]

10.6 PRIME PHASE DIPOLES

Although a link has been established between the spatial-temporal geometry of Pascal's Triangle and the spatial-temporal geometry of the 3-4-5 triangle together with its associated continuous prime proportions, the necessary link between cuboctahedral dipoles and the spatial-temporal prime-phase symmetry of Pythagorean triples remains to be established. This link is required to explain how the dipole field states of the electron, for instance, could possibly be mediated (triangulated) by spatial-temporal triples geometry.

My intent here is not to show the particulars of a specific atomic electron transition from one orbital radius to another. The possible triangulations between radii are evident from the continuous proportions detailed in this chapter. However the energy differences between specific dipole field states are greatly influenced by the symmetry breaking of

[*] δ_e=2: r_v series is 2, 12, 30, 56, ... is 2x the δ_o=1 r_d series of **1**, 6, 15, 28, ... when the **in-circle** is included.

[†] Seemingly in conformance with their relative orientations, notice that δ_o=1 continuous prime proportions are primary phase oriented since they contain 3 primary and one secondary phase polarity, and those of δ_e=2 are secondary phase oriented since they contain 3 secondary and one primary phase polarity.

dimensional energy states within matter. And, although it is possible that such energy differences might ultimately be predicted, at present no such model exists. The specifics of these atomic dipole transitions are interesting to consider, but the more compelling question to answer first is whether or not triples geometry has the capacity to mediate dipole field state transitions in the first place. Without an established symmetric link that allows the two geometries to interface so as to add or subtract momentum from the electron's dipole field there is no hope of ever developing a dimensional number model for these state transitions, however compelling the dimensional geometry of triples might be. I want to know whether or not the spatial-temporal geometry of the primary and secondary prime phases is limited to Pythagorean triples or whether it extends also to the spatial-temporal geometry of the cuboctahedral dipole fields discussed previously.

Thankfully this is a simple question to answer because all it takes is a prime phase analysis similar to Figure 49, which demonstrated that all diagonal terms belong to either the primary or secondary prime phase (6n±1). I have analyzed the Fibonacci and Lucas series in the same way, as shown in Figure 54. From this figure a clear prime phase pattern is evident in both cuboctahedral dipole fields. The dipole field intervals that are responsible for spatial-temporal inversion are always prime-phase terms. They are the two symmetry states on either side of neutral inversion centers and one state on either side of the central spatial-temporal octahedral inversion centers. Notice that the Fibonacci series harbors an octahedral inversion between the prime-phase complements 5 and 13, and that the Lucas series harbors an octahedral inversion between the prime-phase complements 7 and 11. Both these pairs of prime complements sum to the same neutral inversion center (18) of the Lucas series through separate relationships detailed previously in equations 8.2.5 and 8.2.4 respectively.

Returning briefly to the topic of *true* primes discussed in conjunction with Figure 48, notice that the prime phase field of the Fibonacci and Lucas series confirms what I said about 2 and 3 not belonging to either prime phase. In both cases the numbers 2 and 3 are not in prime phase positions in the analog pattern. Yet the number 1 is in a prime phase position in both cases, and it meets the criteria of belonging to the prime phase per Figure 48 since it is 6x0+1.

The cubic poles about which octahedral rotation occurs always belong to the prime phase. This suggests a means of rotational coupling between cuboctahedral dipoles and triples prime geometry. Furthermore, the perspective of the in-circle of the 3-4-5 triangle marks the first merger of spatial and temporal translational perspectives, the last solely singular states, and the last states in the primordial field from the 2:1 spatial/temporal perspective.

Fibonacci: E_I	$\|E_I\|$	$(E_I+1)/6$	$(E_I-1)/6$	Lucas: F_I	$\|F_I\|$	$(F_I+1)/6$	$(F_I-1)/6$
0	9	1/6	- 1/6	2	2	1/2	1/6
1	1	1/3	**0**	**1**	1	1/3	**0**
1	1	1/3	**0**	3	3	2/3	1/3
2	2	1/2	1/6	4	4	5/6	1/2
3	3	2/3	1/3	**7**	7	1 1/3	**1**
5	5	**1**	2/3	**11**	2	**2**	1 2/3
8	8	1 1/2	1 1/6	18	9	3 1/6	2 5/6
13	4	2 1/3	**2**	**29**	2	**5**	4 2/3
21	3	3 2/3	3 1/3	**47**	2	**8**	7 2/3
34	7	5 5/6	5 1/2	76	4	12 5/6	12 1/2
55	1	9 1/3	**9**	123	6	20 2/3	20 1/3
89	8	**15**	14 2/3	**199**	1	33 1/3	**33**
144	9	24 1/6	23 5/6	322	7	53 5/6	53 1/2
233	8	**39**	38 2/3	**521**	8	**87**	86 2/3
377	8	**63**	62 2/3	843	6	140 2/3	140 1/3
610	7	101 5/6	101 1/2	1364	5	227 1/2	227 1/6
987	6	164 2/3	164 1/3	**2207**	2	**368**	367 2/3
1597	4	266 1/3	**266**	**3571**	7	595 1/3	**595**
2584	1	430 5/6	430 1/2	5778	9	963 1/6	962 5/6
4181	5	**697**	696 2/3	**9349**	7	1558 1/3	**1558**
6765	6	1127 2/3	1127 1/3	**15127**	7	2521 1/3	**2521**
10946	2	1824 1/2	1824 1/6	24476	5	4079 1/2	4079 1/6
17711	8	**2952**	2951 2/3	39603	3	6600 2/3	6600 1/3
28657	1	4776 1/3	**4776**	**64079**	8	**10680**	10679 2/3
46368	9	7728 1/6	7727 5/6	103682	2	17280 1/2	17280 1/6
75025	1	12504 1/3	**12504**	**167761**	1	27960 1/3	**27960**
121393	1	20232 1/3	**20232**	271443	3	45240 2/3	45240 1/3
196418	2	32736 1/2	32736 1/6	439204	4	73200 5/6	73200 1/2
317811	3	52968 2/3	52968 1/3	**710647**	7	118441 1/3	**118441**
514229	5	**85705**	85704 2/3	**1149851**	2	**191642**	191641 2/3
832040	8	138673 1/2	138673 1/6	1860498	9	310083 1/6	310082 5/6
1346269	4	224378 1/3	**224378**	**3010349**	2	**501725**	501724 2/3
2178309	3	363051 2/3	363051 1/3	**4870847**	2	**811808**	811807 2/3
3524578	7	587429 5/6	587429 1/2	7881196	4	1313532 5/6	1313532 1/2
5702887	1	950481 1/3	**950481**	12752043	6	2125340 2/3	2125340 1/3
9227465	8	**1537911**	1537910 2/3	**20633239**	1	3438873 1/3	**3438873**
14930352	9	2488392 1/6	2488391 5/6	33385282	7	5564213 5/6	5564213 1/2
24157817	8	**4026303**	4026302 2/3	**54018521**	8	**9003087**	9003086 2/3
39088169	8	**6514695**	6514694 2/3	87403803	6	14567300 2/3	14567300 1/3

Figure 54. Prime phase analysis of the Fibonacci and Lucas series: Primary prime phase terms are highlighted with bold-white and secondary with bold-black. Actual prime terms are underlined. The analysis shows a clear prime-phase pattern for both cuboctahedral dipole fields. This pattern suggests that the prime phase mediates inversion between spatial and temporal polarization.

Therefore, it makes sense to me above other choices that electromagnetism would evolve from this primordial dimensional symmetry. When you add in the nested spatial/temporal continuous proportions of primes, a compelling argument emerges for the symmetry of triples to be a significant player within the dimensional continuum. Moving beyond theory, however, to prove that prime-phase symmetry governs electromagnetism will require some form of experimental corroboration that is beyond the scope of this work. So I'll simply remind the reader of what I said at the onset of this chapter regarding some of the key symmetry elements of electromagnetism and how they relate to the findings in this chapter:

First, there are the orthogonal waveforms of electric and magnetic fields, which the orthogonal spatial and temporal continuous proportions of triples and the spatial and temporal components of the continuous ip^2 proportions both have the capacity to sustain.

Second, there are the discrete indivisible frequencies or wavelengths that I find analogous to primes and the characteristics of interleaving ip^2 proportions, as well as to the general behavior of continuous whole number proportions.

Third, there is complementary color polarization that has been linked to complementary number symmetry of primary and secondary prime phases, which has been suggested to couple with complementary-cuboctahedral super-symmetry and the complementary-cubic 3-6 octahedral dipole rotation axis, since both are related to the same triangular dimensional number symmetry.

Fourth, there is both additive and subtractive coupling between electrons (cuboctahedral dipoles) and photons that accompanies field state transitions. So there is an inherent requirement that the two phenomena be related to the same underlying dimensional symmetry (a *unified field*) such that they can interact as observed – a relationship that has been confirmed by the preceding analysis. Based on quantum theory, the symmetric mechanism for this coupling involves adding or subtracting angular momentum from a given dipole, which means speeding up or slowing down the spin of the dipole on its own rotation axis. This model appears to be consistent with the dimensional symmetry of cuboctahedral dipoles. However dimensional number theory is not developed sufficiently to move beyond conjecture regarding this interaction.

Last, there is the element of quantum theory that suggests that the photon is its own anti-particle (unlike cuboctahedral dipoles), which I believe can be explained together by the bi-directional ip^2 proportion and the dual spatial/temporal (δ_o=1) and temporal/spatial (δ_e=2) perspectives and their converse for the anti-field perspectives.

CHAPTER CITATIONS

[82] D. Zagier *The first 50 million prime numbers* The Mathematical Intelligencer (1977) 7-19
[83] Ore, Oystein 1948 *Number Theory and Its History* McGraw-Hill Book Co. (USA) p 166
[84] Iamblichus trans. Waterfield, R. 1988 *The Theology of Arithmetic* Phanes Press (USA) p 75
[85] Ore, Oystein 1948 *Number Theory and Its History* McGraw-Hill Book Co. (USA) p 92
[86] Ore, Oystein 1948 *Number Theory and Its History* McGraw-Hill Book Co. (USA) p 72

EPILOGUE

The dimensional number theory presented in the preceding chapters was generally accompanied with significant mathematical and geometric proofs. I've provided these proofs as self-evident mathematical truths that can be easily corroborated by independent investigators. Certainly it is possible and perhaps even likely that some of the interpretations I've made in the presentation of this theory will be the subject of controversy, especially since I've suggested that many long-standing entrenched models should be superseded with what I feel is a much more rational and defensible interpretation of the data. Clearly the accuracy of my interpretations rests on my shoulders. However, regardless of whether or not a particular interpretation of the numeric symmetry is wholly accurate, the complementary geometric symmetry of number is self-evident and compelling.

It is my philosophic opinion that the absolute axiomatic theory of everything must ultimately explain why everything is as it is concurrently – the answer lies in the underlying unity connecting all. Therefore, the simplest answer that connects the most dots (greatest breadth) comes closest to being the universal truth underlying all – the mathematical solution that best adheres to the principal of *Occam's Razor*. By connecting the spatial/temporal dimensional geometry of continuous proportions with complementary modulus-9 symmetry transformations, prime number symmetry, quantum number and sub-atomic particle symmetries the dimensional number theory of everything I've put forward has the capacity to be this universal truth.

I've provided numerous examples connecting continuous proportions with ancient number theory and philosophy to reinforce the long-standing notion that everything is ultimately connected thru one common origin: the *Source* or *Monad*, the *Limited-odd* (*Peras*) and *Unlimited-even (Apeiron)* – the germinal symmetry elements of the finite point and the infinite void. In so doing I have provided some of the necessary foundational elements that I hope can help to re-marry scientific and philosophic disciplines through recognition of their common origin – the ancient number theory of the Pythagoreans and the near eastern axis-age melting pot of ideas that informed them.

The fundamental natural harmony of continuous proportions and their complementary modulus-9 symmetry transformations has the capacity to completely change the way mathematical concepts are understood. It seems likely that significant advancements in number theory will come from treating mathematical operations as complementary symmetry transformations because it is essentially a new way of interpreting mathematics. For instance, *Amicable Numbers* — composed of each other's *Aliquot parts* (divisors) — yet another expression of complementary numeric symmetry:

Complementary (Amicable) Numbers

220	4	1184	5	2620	1	5020	7	6232	4	10744	7	17296	7	63020	2
284	5	1210	4	2924	8	5564	2	6368	5	10856	2	18416	2	76084	7

Imagine navigating polynomial equations through a combination of complementary translational and rotational symmetry transformations. Imagine a new complex number system based on modulus-9 spatial-temporal dimensional number symmetry. Imagine *quantum computing* with nine complementary numbers versus two binary. Right now the applications of modulus-9 are for the most part unknown — leaving the field wide open for new exploration.

I've offered up a logical means of moving past the theoretical barrier quantum physics has run up against. Matter is a complex medium, but with time I can envision dimensional number theory unlocking many of its remaining mysteries. For the most part I've steered clear of astrophysics in this book, although I've essentially begun the process of merging quantum theory and relativity into a transcendent *unified field theory*. Dimensional number theory has the capacity to revolutionize the way theorists envision the origin and history of the universe.

The body of experimental findings in quantum physics essentially corroborates the dimensional number theory I've proposed. I've intentionally taken a qualitative approach to describing the symmetric forces that govern matter because I think answering the question why is far more powerful than simply giving *predictions that agree with experiment* in the long run. Knowing why should ultimately lead to better experimental predictability, but work is required to understand the energetics of symmetry breaking before this will be possible. Applying the golden proportion to the present condition, one can predict a merging of past and present theory to inform the evolving future absolute theory of everything,

Needless to say, the upsides of a unified field theory for scientific advancements are huge. However there are some serious hurdles to overcome before such advancements are even conceivable. Given the entrenchment of quantum mechanics and relativity within the scientific establishment, there are significant barriers and counter incentives to actually reaching a conclusive answer. That being said, the dimensional number theory I've presented in this book is compelling, and I find it hard to imagine what kind of rational counter argument could be mounted against it. So I think it's only a matter of time before dimensions of energy are understood through geometric axioms of continuous proportions and their accompanying complementary symmetry transformations. I'm hopeful that this future turning point will mark the beginning of an exciting new era of philosophical scientific enlightenment.

My study of continuous proportion has given me new insights that extend far beyond the scope of this book. Ultimately if all-that-is can be traced back to one singular energetic dimensional origin. So it's interesting to consider the possibility that all life might be complex forms of continuous proportion perpetuated by this complementary dimensional energy symmetry. I see the continuous golden proportion as a compelling archetypal symmetry for evolution. Passing this notion through my own filter, I see that, like continuous golden proportion, I leverage the sum total of my past experiences to respond to the present condition thereby continuously shaping my future (past + present = future), although certainly not with such mathematical precision.

When I look at the singular beginning of every life form (seeds, eggs and cells) I can't help but think of the singular spherical point at the origin of every dimension. The possibly that all forms of life evolve from the same singular dimensional origin of energy is intriguing. Given that the continuous golden proportion *plays an eminent role in Botany* and *plays a dominating part in the proportions of the human body*, I think it's reasonable to expect genetics to be somehow connected to this archetypal dimensional symmetry. So I'm hopeful that we will ultimately be able to place evolution within the context of the dimensional continuum – through life the universe strives to become self-aware.

Appendix A1. Triples tables: fixed d_o (1, 9, 25) on the left and fixed d_e (2, 8, 18) on the right, the first 18 in-radii for each series is provided, primes underlined.

incircle	excircles			triples		○p	△p	differences	
r	r_v	r_h	r_d	v	h	d	$v+h$	δ_o	δ_e
1	2	3	6	3	4	5	7	1	2
2	3	10	15	5	12	13	17	1	8
3	4	21	28	7	24	25	31	1	18
4	5	36	45	9	40	41	49	1	32
5	6	55	66	11	60	61	71	1	50
6	7	78	91	13	84	85	97	1	72
7	8	105	120	15	112	113	127	1	98
8	9	136	153	17	144	145	161	1	128
9	10	171	190	19	180	181	199	1	162
10	11	210	231	21	220	221	241	1	200
11	12	253	276	23	264	265	287	1	242
12	13	300	325	25	312	313	337	1	288
13	14	351	378	27	364	365	391	1	338
14	15	406	435	29	420	421	449	1	392
15	16	465	496	31	480	481	511	1	450
16	17	528	561	33	544	545	577	1	512
17	18	595	630	35	612	613	647	1	578
18	19	666	703	37	684	685	721	1	648

incircle	excircles			triples		○p	△p	differences	
r	r_v	r_h	r_d	v	h	d	$v+h$	δ_o	δ_e
1	2	3	6	3	4	5	7	1	2
3	12	5	20	15	8	17	23	9	2
5	30	7	42	35	12	37	47	25	2
7	56	9	72	63	16	65	79	49	2
9	90	11	110	99	20	101	119	81	2
11	132	13	156	143	24	145	167	121	2
13	182	15	210	195	28	197	223	169	2
15	240	17	272	255	32	257	287	225	2
17	306	19	342	323	36	325	359	289	2
19	380	21	420	399	40	401	439	361	2
21	462	23	506	483	44	485	527	441	2
23	552	25	600	575	48	577	623	529	2
25	650	27	702	675	52	677	727	625	2
27	756	29	812	783	56	785	839	729	2
29	870	31	930	899	60	901	959	841	2
31	992	33	1056	1023	64	1025	1087	961	2
33	1122	35	1190	1155	68	1157	1223	1089	2
35	1260	37	1332	1295	72	1297	1367	1225	2

incircle	excircles			triples		○p	△p	differences	
r	r_v	r_h	r_d	v	h	d	$v+h$	δ_o	δ_e
3	12	5	20	15	8	17	23	9	2
6	15	14	35	21	20	29	41	9	8
9	18	27	54	27	36	45	63	9	18
12	21	44	77	33	56	65	89	9	32
15	24	65	104	39	80	89	119	9	50
18	27	90	135	45	108	117	153	9	72
21	30	119	170	51	140	149	191	9	98
24	33	152	209	57	176	185	233	9	128
27	36	189	252	63	216	225	279	9	162
30	39	230	299	69	260	269	329	9	200
33	42	275	350	75	308	317	383	9	242
36	45	324	405	81	360	369	441	9	288
39	48	377	464	87	416	425	503	9	338
42	51	434	527	93	476	485	569	9	392
45	54	495	594	99	540	549	639	9	450
48	57	560	665	105	608	617	713	9	512
51	60	629	740	111	680	689	791	9	578
54	63	702	819	117	756	765	873	9	648

incircle	excircles			triples		○p	△p	differences	
r	r_v	r_h	r_d	v	h	d	$v+h$	δ_o	δ_e
2	3	10	15	5	12	13	17	1	8
6	15	14	35	21	20	29	41	9	8
10	35	18	63	45	28	53	73	25	8
14	63	22	99	77	36	85	113	49	8
18	99	26	143	117	44	125	161	81	8
22	143	30	195	165	52	173	217	121	8
26	195	34	255	221	60	229	281	169	8
30	255	38	323	285	68	293	353	225	8
34	323	42	399	357	76	365	433	289	8
38	399	46	483	437	84	445	521	361	8
42	483	50	575	525	92	533	617	441	8
46	575	54	675	621	100	629	721	529	8
50	675	58	783	725	108	733	833	625	8
54	783	62	899	837	116	845	953	729	8
58	899	66	1023	957	124	965	1081	841	8
62	1023	70	1155	1085	132	1093	1217	961	8
66	1155	74	1295	1221	140	1229	1361	1089	8
70	1295	78	1443	1365	148	1373	1513	1225	8

incircle	excircles			triples		○p	△p	differences	
r	r_v	r_h	r_d	v	h	d	$v+h$	δ_o	δ_e
5	30	7	42	35	12	37	47	25	2
10	35	18	63	45	28	53	73	25	8
15	40	33	88	55	48	73	103	25	18
20	45	52	117	65	72	97	137	25	32
25	50	75	150	75	100	125	175	25	50
30	55	102	187	85	132	157	217	25	72
35	60	133	228	95	168	193	263	25	98
40	65	168	273	105	208	233	313	25	128
45	70	207	322	115	252	277	367	25	162
50	75	250	375	125	300	325	425	25	200
55	80	297	432	135	352	377	487	25	242
60	85	348	493	145	408	433	553	25	288
65	90	403	558	155	468	493	623	25	338
70	95	462	627	165	532	557	697	25	392
75	100	525	700	175	600	625	775	25	450
80	105	592	777	185	672	697	857	25	512
85	110	663	858	195	748	773	943	25	578
90	115	738	943	205	828	853	1033	25	648

incircle	excircles			triples		○p	△p	differences	
r	r_v	r_h	r_d	v	h	d	$v+h$	δ_o	δ_e
3	4	21	28	7	24	25	31	1	18
9	18	27	54	27	36	45	63	9	18
15	40	33	88	55	48	73	103	25	18
21	70	39	130	91	60	109	151	49	18
27	108	45	180	135	72	153	207	81	18
33	154	51	238	187	84	205	271	121	18
39	208	57	304	247	96	265	343	169	18
45	270	63	378	315	108	333	423	225	18
51	340	69	460	391	120	409	511	289	18
57	418	75	550	475	132	493	607	361	18
63	504	81	648	567	144	585	711	441	18
69	598	87	754	667	156	685	823	529	18
75	700	93	868	775	168	793	943	625	18
81	810	99	990	891	180	909	1071	729	18
87	928	105	1120	1015	192	1033	1207	841	18
93	1054	111	1258	1147	204	1165	1351	961	18
99	1188	117	1404	1287	216	1305	1503	1089	18
105	1330	123	1558	1435	228	1453	1663	1225	18

Appendix A2. Triples tables: fixed d_o (49, 81, 121) on the left and fixed d_e (32, 50, 72) on the right, the first 18 in-radii for each series is provided, primes underlined.

incircle	excircles			triples		○p	△p	differences	
r	r_v	r_h	r_d	v	h	d	$v+h$	δ_o	δ_e
7	56	9	72	63	16	65	79	49	2
14	63	22	99	77	36	85	113	49	8
21	70	39	130	91	60	109	151	49	18
28	77	60	165	105	88	137	193	49	32
35	84	85	204	119	120	169	239	49	50
42	91	114	247	133	156	205	289	49	72
49	98	147	294	147	196	245	343	49	98
56	105	184	345	161	240	289	401	49	128
63	112	225	400	175	288	337	463	49	162
70	119	270	459	189	340	389	529	49	200
77	126	319	522	203	396	445	599	49	242
84	133	372	589	217	456	505	673	49	288
91	140	429	660	231	520	569	751	49	338
98	147	490	735	245	588	637	833	49	392
105	154	555	814	259	660	709	919	49	450
112	161	624	897	273	736	785	1009	49	512
119	168	697	984	287	816	865	1103	49	578
126	175	774	1075	301	900	949	1201	49	648

incircle	excircles			triples		○p	△p	differences	
r	r_v	r_h	r_d	v	h	d	$v+h$	δ_o	δ_e
4	5	36	45	9	40	41	49	1	32
12	21	44	77	33	56	65	89	9	32
20	45	52	117	65	72	97	137	25	32
28	77	60	165	105	88	137	193	49	32
36	117	68	221	153	104	185	257	81	32
44	165	76	285	209	120	241	329	121	32
52	221	84	357	273	136	305	409	169	32
60	285	92	437	345	152	377	497	225	32
68	357	100	525	425	168	457	593	289	32
76	437	108	621	513	184	545	697	361	32
84	525	116	725	609	200	641	809	441	32
92	621	124	837	713	216	745	929	529	32
100	725	132	957	825	232	857	1057	625	32
108	837	140	1085	945	248	977	1193	729	32
116	957	148	1221	1073	264	1105	1337	841	32
124	1085	156	1365	1209	280	1241	1489	961	32
132	1221	164	1517	1353	296	1385	1649	1089	32
140	1365	172	1677	1505	312	1537	1817	1225	32

incircle	excircles			triples		○p	△p	differences	
r	r_v	r_h	r_d	v	h	d	$v+h$	δ_o	δ_e
9	90	11	110	99	20	101	119	81	2
18	99	26	143	117	44	125	161	81	8
27	108	45	180	135	72	153	207	81	18
36	117	68	221	153	104	185	257	81	32
45	126	95	266	171	140	221	311	81	50
54	135	126	315	189	180	261	369	81	72
63	144	161	368	207	224	305	431	81	98
72	153	200	425	225	272	353	497	81	128
81	162	243	486	243	324	405	567	81	162
90	171	290	551	261	380	461	641	81	200
99	180	341	620	279	440	521	719	81	242
108	189	396	693	297	504	585	801	81	288
117	198	455	770	315	572	653	887	81	338
126	207	518	851	333	644	725	977	81	392
135	216	585	936	351	720	801	1071	81	450
144	225	656	1025	369	800	881	1169	81	512
153	234	731	1118	387	884	965	1271	81	578
162	243	810	1215	405	972	1053	1377	81	648

incircle	excircles			triples		○p	△p	differences	
r	r_v	r_h	r_d	v	h	d	$v+h$	δ_o	δ_e
5	6	55	66	11	60	61	71	1	50
15	24	65	104	39	80	89	119	9	50
25	50	75	150	75	100	125	175	25	50
35	84	85	204	119	120	169	239	49	50
45	126	95	266	171	140	221	311	81	50
55	176	105	336	231	160	281	391	121	50
65	234	115	414	299	180	349	479	169	50
75	300	125	500	375	200	425	575	225	50
85	374	135	594	459	220	509	679	289	50
95	456	145	696	551	240	601	791	361	50
105	546	155	806	651	260	701	911	441	50
115	644	165	924	759	280	809	1039	529	50
125	750	175	1050	875	300	925	1175	625	50
135	864	185	1184	999	320	1049	1319	729	50
145	986	195	1326	1131	340	1181	1471	841	50
155	1116	205	1476	1271	360	1321	1631	961	50
165	1254	215	1634	1419	380	1469	1799	1089	50
175	1400	225	1800	1575	400	1625	1975	1225	50

incircle	excircles			triples		○p	△p	differences	
r	r_v	r_h	r_d	v	h	d	$v+h$	δ_o	δ_e
11	132	13	156	143	24	145	167	121	2
22	143	30	195	165	52	173	217	121	8
33	154	51	238	187	84	205	271	121	18
44	165	76	285	209	120	241	329	121	32
55	176	105	336	231	160	281	391	121	50
66	187	138	391	253	204	325	457	121	72
77	198	175	450	275	252	373	527	121	98
88	209	216	513	297	304	425	601	121	128
99	220	261	580	319	360	481	679	121	162
110	231	310	651	341	420	541	761	121	200
121	242	363	726	363	484	605	847	121	242
132	253	420	805	385	552	673	937	121	288
143	264	481	888	407	624	745	1031	121	338
154	275	546	975	429	700	821	1129	121	392
165	286	615	1066	451	780	901	1231	121	450
176	297	688	1161	473	864	985	1337	121	512
187	308	765	1260	495	952	1073	1447	121	578
198	319	846	1363	517	1044	1165	1561	121	648

incircle	excircles			triples		○p	△p	differences	
r	r_v	r_h	r_d	v	h	d	$v+h$	δ_o	δ_e
6	7	78	91	13	84	85	97	1	72
18	27	90	135	45	108	117	153	9	72
30	55	102	187	85	132	157	217	25	72
42	91	114	247	133	156	205	289	49	72
54	135	126	315	189	180	261	369	81	72
66	187	138	391	253	204	325	457	121	72
78	247	150	475	325	228	397	553	169	72
90	315	162	567	405	252	477	657	225	72
102	391	174	667	493	276	565	769	289	72
114	475	186	775	589	300	661	889	361	72
126	567	198	891	693	324	765	1017	441	72
138	667	210	1015	805	348	877	1153	529	72
150	775	222	1147	925	372	997	1297	625	72
162	891	234	1287	1053	396	1125	1449	729	72
174	1015	246	1435	1189	420	1261	1609	841	72
186	1147	258	1591	1333	444	1405	1777	961	72
198	1287	270	1755	1485	468	1557	1953	1089	72
210	1435	282	1927	1645	492	1717	2137	1225	72

Appendix A3. Triples tables: fixed d_o (169, 225, 289) on the left and fixed d_e (98, 128, 162) on the right, the first 18 in-radii for each series is provided, primes underlined.

incircle	excircles			triples		○p	Δp	differences	
r	r_v	r_h	r_d	v	h	d	$v+h$	δ_o	δ_e
13	182	15	210	195	28	197	223	169	2
26	195	34	255	221	60	229	281	169	8
39	208	57	304	247	96	265	343	169	18
52	221	84	357	273	136	305	409	169	32
65	234	115	414	299	180	349	479	169	50
78	247	150	475	325	228	397	553	169	72
91	260	189	540	351	280	449	631	169	98
104	273	232	609	377	336	505	713	169	128
117	286	279	682	403	396	565	799	169	162
130	299	330	759	429	460	629	889	169	200
143	312	385	840	455	528	697	983	169	242
156	325	444	925	481	600	769	1081	169	288
169	338	507	1014	507	676	845	1183	169	338
182	351	574	1107	533	756	925	1289	169	392
195	364	645	1204	559	840	1009	1399	169	450
208	377	720	1305	585	928	1097	1513	169	512
221	390	799	1410	611	1020	1189	1631	169	578
234	403	882	1519	637	1116	1285	1753	169	648

incircle	excircles			triples		○p	Δp	differences	
r	r_v	r_h	r_d	v	h	d	$v+h$	δ_o	δ_e
7	8	105	120	15	112	113	127	1	98
21	30	119	170	51	140	149	191	9	98
35	60	133	228	95	168	193	263	25	98
49	98	147	294	147	196	245	343	49	98
63	144	161	368	207	224	305	431	81	98
77	198	175	450	275	252	373	527	121	98
91	260	189	540	351	280	449	631	169	98
105	330	203	638	435	308	533	743	225	98
119	408	217	744	527	336	625	863	289	98
133	494	231	858	627	364	725	991	361	98
147	588	245	980	735	392	833	1127	441	98
161	690	259	1110	851	420	949	1271	529	98
175	800	273	1248	975	448	1073	1423	625	98
189	918	287	1394	1107	476	1205	1583	729	98
203	1044	301	1548	1247	504	1345	1751	841	98
217	1178	315	1710	1395	532	1493	1927	961	98
231	1320	329	1880	1551	560	1649	2111	1089	98
245	1470	343	2058	1715	588	1813	2303	1225	98

incircle	excircles			triples		○p	Δp	differences	
r	r_v	r_h	r_d	v	h	d	$v+h$	δ_o	δ_e
15	240	17	272	255	32	257	287	225	2
30	255	38	323	285	68	293	353	225	8
45	270	63	378	315	108	333	423	225	18
60	285	92	437	345	152	377	497	225	32
75	300	125	500	375	200	425	575	225	50
90	315	162	567	405	252	477	657	225	72
105	330	203	638	435	308	533	743	225	98
120	345	248	713	465	368	593	833	225	128
135	360	297	792	495	432	657	927	225	162
150	375	350	875	525	500	725	1025	225	200
165	390	407	962	555	572	797	1127	225	242
180	405	468	1053	585	648	873	1233	225	288
195	420	533	1148	615	728	953	1343	225	338
210	435	602	1247	645	812	1037	1457	225	392
225	450	675	1350	675	900	1125	1575	225	450
240	465	752	1457	705	992	1217	1697	225	512
255	480	833	1568	735	1088	1313	1823	225	578
270	495	918	1683	765	1188	1413	1953	225	648

incircle	excircles			triples		○p	Δp	differences	
r	r_v	r_h	r_d	v	h	d	$v+h$	δ_o	δ_e
8	9	136	153	17	144	145	161	1	128
24	33	152	209	57	176	185	233	9	128
40	65	168	273	105	208	233	313	25	128
56	105	184	345	161	240	289	401	49	128
72	153	200	425	225	272	353	497	81	128
88	209	216	513	297	304	425	601	121	128
104	273	232	609	377	336	505	713	169	128
120	345	248	713	465	368	593	833	225	128
136	425	264	825	561	400	689	961	289	128
152	513	280	945	665	432	793	1097	361	128
168	609	296	1073	777	464	905	1241	441	128
184	713	312	1209	897	496	1025	1393	529	128
200	825	328	1353	1025	528	1153	1553	625	128
216	945	344	1505	1161	560	1289	1721	729	128
232	1073	360	1665	1305	592	1433	1897	841	128
248	1209	376	1833	1457	624	1585	2081	961	128
264	1353	392	2009	1617	656	1745	2273	1089	128
280	1505	408	2193	1785	688	1913	2473	1225	128

incircle	excircles			triples		○p	Δp	differences	
r	r_v	r_h	r_d	v	h	d	$v+h$	δ_o	δ_e
17	306	19	342	323	36	325	359	289	2
34	323	42	399	357	76	365	433	289	8
51	340	69	460	391	120	409	511	289	18
68	357	100	525	425	168	457	593	289	32
85	374	135	594	459	220	509	679	289	50
102	391	174	667	493	276	565	769	289	72
119	408	217	744	527	336	625	863	289	98
136	425	264	825	561	400	689	961	289	128
153	442	315	910	595	468	757	1063	289	162
170	459	370	999	629	540	829	1169	289	200
187	476	429	1092	663	616	905	1279	289	242
204	493	492	1189	697	696	985	1393	289	288
221	510	559	1290	731	780	1069	1511	289	338
238	527	630	1395	765	868	1157	1633	289	392
255	544	705	1504	799	960	1249	1759	289	450
272	561	784	1617	833	1056	1345	1889	289	512
289	578	867	1734	867	1156	1445	2023	289	578
306	595	954	1855	901	1260	1549	2161	289	648

incircle	excircles			triples		○p	Δp	differences	
r	r_v	r_h	r_d	v	h	d	$v+h$	δ_o	δ_e
9	10	171	190	19	180	181	199	1	162
27	36	189	252	63	216	225	279	9	162
45	70	207	322	115	252	277	367	25	162
63	112	225	400	175	288	337	463	49	162
81	162	243	486	243	324	405	567	81	162
99	220	261	580	319	360	481	679	121	162
117	286	279	682	403	396	565	799	169	162
135	360	297	792	495	432	657	927	225	162
153	442	315	910	595	468	757	1063	289	162
171	532	333	1036	703	504	865	1207	361	162
189	630	351	1170	819	540	981	1359	441	162
207	736	369	1312	943	576	1105	1519	529	162
225	850	387	1462	1075	612	1237	1687	625	162
243	972	405	1620	1215	648	1377	1863	729	162
261	1102	423	1786	1363	684	1525	2047	841	162
279	1240	441	1960	1519	720	1681	2239	961	162
297	1386	459	2142	1683	756	1845	2439	1089	162
315	1540	477	2332	1855	792	2017	2647	1225	162

Appendix

Appendix B1. Spatially polarized Complementary Cuboctahedron template: For best results color photocopy template on to card stock. Cut out and then crease along edges of adjoining faces and assembly tabs. Then glue and secure tabs until adhesive dries.

Appendix B2. Temporally polarized Complementary Cuboctahedron template: For best results color photocopy template on to card stock. Cut out and then crease along edges of adjoining faces and assembly tabs. Then glue and secure tabs until adhesive dries.

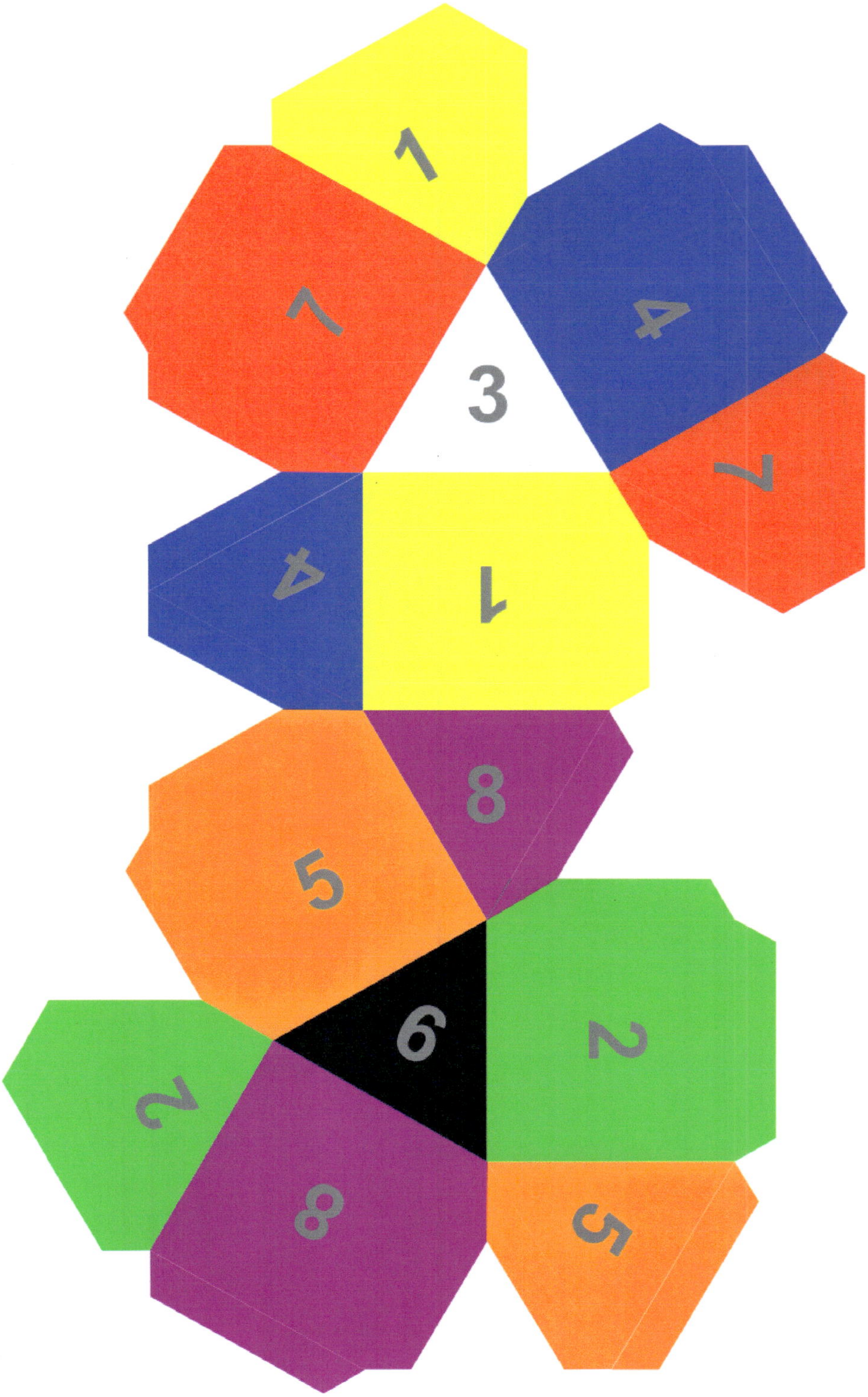

About the Author

Peter Williams is a materials scientist, engineer, and a pioneer in dimensional number theory. He lives in Portland, Oregon. This is his first book.

Learn more at: **www.naturalnumber.com**

Made in the USA
Monee, IL
01 May 2026

49348869R00114